Urban Gardening

Learn Step-By-Step How To Grow In Containers And Everything About Balcony And Vertical Gardening. Build Your Own Garden in Any City Apartment

By

Matt Mitchell

© **Copyright 2020 by Matt Mitchell - All rights reserved.**

This document is geared towards providing exact and reliable information in regards to the topic and issue covered. The publication is sold with the idea that the publisher is not required to render accounting, officially permitted, or otherwise, qualified services. If advice is necessary, legal or professional, a practiced individual in the profession should be ordered.

- From a Declaration of Principles which was accepted and approved equally by a Committee of the American Bar Association and a Committee of Publishers and Associations.

In no way is it legal to reproduce, duplicate, or transmit any part of this document in either electronic means or in printed format. Recording of this publication is strictly prohibited and any storage of this document is not allowed unless with written permission from the publisher. All rights reserved.

The information provided herein is stated to be truthful and consistent, in that any liability, in terms of inattention or otherwise, by any usage or abuse of any policies, processes, or directions contained within is the solitary and utter responsibility of the recipient reader.

Under no circumstances will any legal responsibility or blame be held against the publisher for any reparation, damages, or monetary loss due to the information herein, either directly or indirectly.

Respective authors own all copyrights not held by the publisher.

The information herein is offered for informational purposes solely, and is universal as so. The presentation of the information is without contract or any type of guarantee assurance.

The trademarks that are used are without any consent, and the publication of the trademark is without permission or backing by the trademark owner. All trademarks and brands within this book are for clarifying purposes only and are the owned by the owners themselves, not affiliated with this document.

Table of Contents

INTRODUCTION 6

CHAPTER 1: URBAN GARDENING AND ITS IMPORTANCE 9

1.1: Importance of Urban Gardening 9

1.2: Urban Farming Is the Future 34

1.3 Difference between Urban and Rural Gardening 45

CHAPTER 2: PLANTING A GARDEN IN AN URBAN AREA 57

2.1 Making a Garden in Urban Area 58

2.2 Steps of Planting a Garden 61

2.3 Why is It Difficult to have a Garden in Urban Area? 72

2.4 Planting Flowers and Vegetables Together 74

CHAPTER 3: CONTAINER GARDENING IN URBAN AREAS 81

3.1 What is Urban Container Gardening? 81

3.2 Benefits of Container Gardening 99

3.3 Steps to Grow Plants in Containers 113

3.4 Best Plants to Grow Using Container Gardening... 118

3.5 Maintenance of Container Garden............................ 127

CHAPTER 4: VERTICAL GARDENING IN URBAN AREAS 134

4.1 Purpose of Vertical Gardening 135

4.2 Benefits of Vertical Gardening 141

4.3 Steps to Grow Plants in a Vertical Garden............... 148

4.4 Best Plants to Grow in Vertical Gardening.............. 156

4.5 Maintenance of Vertical Garden 164

CONCLUSION.. 178

REFERENCES.. 179

Introduction

Urban gardening is the method of growing plants in an urban climate, of all styles and varieties. Urban gardening, also known as urban horticulture or urban farming, incorporates many distinct types of gardening, including:

Container gardening:

Popular to people with small patios, yards, or balconies. Container gardening uses a range of containers to cultivate all sorts of plants for food or beauty – pots, old tires, raised beds, window boxes, kiddie pools, barrels, shoes, and watering cans.

Indoor gardening:

If there are no patios, decks, yards, or balconies open, indoor gardening can also be an efficient form of urban gardening. Plants can be grown in container-like containers, as well as in indoor greenhouses or solariums (sunrooms).

Community gardening:

It is a way of using public or private outdoor spaces to grow gardens as a community for food or pleasure and is a great option for those with no yard or outdoor space.

Guerilla gardening:

A more radical type of urban gardening, guerrilla gardening, is a way to introduce plants to public spaces that don't actually belong to the gardener, such as a parking lot, median, next to a highway, or in small stretches of dirt.

Green roofs:

Roofs built for the cultivation of plants with a growing medium are also a method of urban gardening and they can be used to grow fruit, trees, and many other plant types. Urban gardening offers many economic, social, and health benefits:

- Provides a local food source
- Bring neighborhoods and families together.
- Educates your children about the origins of food.
- Adds green spaces to cities
- Helps avoid soil erosion
- Mitigates storm water runoff
- Helps trap air and rainwater
- Reduce urban heat island impact

From the past, what will come to mind when talking about planting and farming were pictures of the fields lined with rows of vegetables? Yet time is changing. Our local food production is rapidly taking place in our urban centers. Nonetheless, about 15 percent of the world's food supply is now produced in urban areas, according to the USDA. And this trend is forecast to expand as the population of the world is projected to become even more of an urban dwelling.

Whether it's by tiny backyard plots, community planting in urban parks, guerrilla planting on vacant lots, indoor hanging gardens, rooftop growing, vertical gardens, and more, urban farming is now a thing.

Urban gardening and farming are mishmashes of growing and raising food techniques and approaches in highly populated urban centers. Because of the very existence of cities, there is not a one-size-fits-all approach, but rather a multitude of approaches and activities pursued by individuals, groups, cooperatives, and companies alike. In an indoor greenhouse, a restaurant may grow its own herbs, a community may take over a vacant lot for an elevated bed greenhouse, a cooperative may hold honey bees on the roof, or a family may design a container garden for a patio — all examples of urban gardening. City gardeners are taking matters into their own hands to grow fresh and organic food instead of the long-standing tradition of trucking in the food to cities.

Chapter 1: Urban Gardening and its Importance

Gardens produce more fine and delicious food than just healthy food. Urban agriculture brings together people with a shared interest - food. Increasing the capacity to build an atmosphere that genuinely sustains the inhabitants improves the overall wellbeing of society.

1.1: Importance of Urban Gardening

Urban farming is important in many respects.

- It is fun for the whole family
- Improves the relations with neighborhood

- Encourage organic farming
- It makes life colorful
- It attracts more birds and bees
- Help the planet
- It helps in good nutrition
- It helps in keeping good health
- It changes the outlook of life
- It saves money
- Helps in controlling Environmental pollution
- It helps in improving physical activity.

Around the same time, the simple act of planting a garden will influence topics such as economics, health, and politics, as food is an important subject of human activity. As the urban agriculture movement rises, here are five ways to turn our environment.

Renewed local economies:

Local neighbor-to-neighbor interaction in our neighborhoods usually doesn't happen. Quick never do residential areas include open spaces where neighborhood exchanges can occur. Likewise, as in most places selling homemade bread to your neighbors is illegal, the law discourages neighborhood exchange, and instead allows you to buy from the grocery chain. The urban farming movement has reinvigorated local exchange within my own city. Instead of buying oranges, from my neighbor's vine, I now swap pumpkin for oranges. If urban farming continues to expand, the introduction of local food production that would compete with the corporate standard on price, quality, convenience, and level of service will cause a significant and positive economic disruption.

Environmental stewardship:

Modern agriculture is a big source of emissions from fossil fuels. Petrochemicals are used for fertilization, spraying, and food protection. The food is filled with oil-based plastics, and fuel is used to transport food worldwide. Urban farming unplugs us from oil by reducing the footprint of transport and using methods of organic growth. While industrial farming often maneuvers to avoid paying for environmental externalities, urban farmers bear the ecological costs of their actions directly. This makes urban farmers better overseers of their land because they derive from it their nutrition. Instead of using toxic substances that kill soil biodiversity, the philosophy of urban farming emphasizes sustainable organic techniques that enrich the topsoil.

A focus on local politics:

Urban agriculture makes it simpler and easier for people to get involved in local politics by getting problems to the forefront that directly impact communities. Local laws are much more important to a person's daily life trying to develop his or her own food than any of the topics usually discussed on CNN.

Urban farming development has also led to large-scale legislative drives such as the California Cottage Food Act, which would allow people to legally sell such homemade products such as jams and pieces of bread. Other community concerns like chickens raising, beekeeping for honey production, or water chlorination are already on the sights of urban farmers and environmentalists alike.

A revolution of health and nutrition:

Improved understanding of the negative health effects of food from the global food chain is a major reason why urban farmers cultivate their own food themselves. When you feed your family's food, you are less likely to douse it in poisons. Local food has more freshness, taste, and preservation of nutrients, so less transportation and processing is involved. When the urban farming movement expands, this will mean greater exposure to local nutritious food and more time spent doing the good physical gardening job. This could lead to less obesity, less chronic illness, and lower spending on health care.

A flowering of community interaction:

Urban farming is a Community-centered lifestyle. After all, growing food is a cooperative effort. I see in my own community that information is exchanged about how and what to grow, seeds are swapped, labor is shared, and the harvest is traded. A stronger interdependence between communities is likely to result as urban farming expands as local food systems introduce more community engagement into the everyday lives of people.

The most important change in our time. While there are many other notable initiatives today, the influence of urban farming is particularly widespread because more people live in towns than rural areas, and food is a basic need that affects all at once. The seeds of progress are being planted around the world now in homes like mine. To grow and blossom these seeds, we need to demand more local food so that the urban-grown growing sector grows. We do need to put pressure to bear on our legal system to promote local trade and more local food production.

Imagine growing fruit, rather than grass. Every society is a local food economy waiting for life to come. The solution to climate change, the health crisis, and the economic recession is right outside your door. I'll meet you at a fence in the backyard.

Importance:

For most high school students, watching plants grow isn't a fascinating pastime. Neither does it do the chores, nor goes to school. And there are growing numbers of schools now having a portion of urban agriculture in their curriculum, and hopefully growing the number of young people in the agricultural sector, experts claim. Purdue University Extension educators describe urban agriculture as simply growing or processing food in urban spaces. Urban agriculture comes in many types, but urban farms, community gardens, and hydroponics or aquaponics programs are the most common. Urban farming projects will assist local communities in both an economic and social way. They allow people to relate to their food more quickly, as well as help stimulate a local economy. Urban farming initiatives such as community gardens may target young people in contexts of non-traditional agriculture.

Here are some ways in which urban farms help their communities improve their health:

REDUCE CARBON EMISSIONS

By finding produce, urban farms are reducing the significant amount of fossil fuel consumption required to store, package, and sell food.

The typical meal just to get to your table has driven 4,200 miles. Urban farming allows customers that their "food print" by giving them the ability to purchase food that has been grown within their neighborhood.

INNOVATIVE TECHNIQUES

As urban areas lack the wide-open fertile grounds of traditional methods of farming, urban farmers are tasked with seeking innovative solutions to problems such as waste, land, capital, and electricity. Of this purpose, more effective technologies are developed to increase the quality and quantity of food which can be provided with the least amount of resources. (For example, The That Experience's vertical aquaponics systems in Long Beach produce 3-4 times as much produce as conventional farming methods and use considerably less water.)

JOB CREATION

From window box herb gardens to open community areas, these farms provide opportunities for community participation. Urban farms establish (and volunteer) job opportunities in big cities, where poverty and hunger are still chronic issues. Small business growth boosts the local economy and benefits the community by generating jobs where people live.

ECONOMIC GROWTH

Because of their proximity to customers, urban farms stimulate the local economy through the redistribution of revenue throughout the city. Farmers are more linked to their customers without a complicated distribution network and able to respond quickly to demand, optimizing income. Furthermore, many of these organizations are organized in a way that offers economic value to the community and helps low-income families by stabilizing food prices and providing subsidized or free goods in many instances.

COMMUNITY BUILDING

Gardens do more than grow good, delicious food. Urban agriculture brings together people with a shared interest-food. Growing the capacity to build an atmosphere that genuinely sustains the inhabitants improves the overall wellbeing of society. Many urban farming ventures involve a high degree of social organization, providing a vested interest to many individuals in the group for their success.

PUBLIC HEALTH

Rising urban communities suffer from obesity and a number of other diet-related health problems.

There are many direct health benefits of introducing healthy food of local communities, including reducing the risk of unhealthy conditions such as heart disease, obesity, diabetes, and more. The presence of individuals in the garden itself offers an opportunity for exercise and a more meaningful connection to farming.

FOOD QUALITY

Smaller scale, local markets provide farmers with the ability to cultivate more specific product varieties. These farms sustain biodiversity by increasing varieties of heirlooms or those with lower shelf-stability. Business proximity and accessibility allows for the introduction of new, healthy goods to populations that have never had access to it in the past.

FOOD SECURITY

While food shortages may not exist in most regions, access issues are absolutely prevalent, especially in urban areas. Around 400,000 people in Orange County don't have access to safe, nutritious food.

Urban agriculture aims to address this by raising the price of nutritious food by removing the middleman and increasing the ability to engage in the growing of this food for community members in need.

Most urban farms often follow charitable models in an attempt to benefit disadvantaged communities through direct contributions or through the distribution of reduced or free produce.

EDUCATION

Urban farming solves another problem that is implicit in our current food culture — a disconnection to where our food comes from. Farmers are raising the sustainability of our future food systems by engaging children and adults alike in education around sustainable, local agriculture.

GREEN SPACE

Eventually, urban agriculture offers something more obvious — greener land. That contributes to a variety of ways to the health of city ecosystems. Greenery adds visual appeal, decreases runoff from precipitation, provides the city with restful spaces, and counteracts the effect of heat-insulating by fixing carbon through photosynthesis.

Benefits of Urban Gardening

Urban agriculture, also known as green farming, is a way for urban dwellers to produce their own food, or have access to local food at least. Urban farming practices are becoming increasingly common in western areas of North America.

With the many advantages of urban agriculture, and all that local food production has to offer, it's important that we keep raising consciousness about how individuals and communities can create a base for better health, social engagement, and economic prosperity.

Here are some of the benefits of urban agriculture:

Increases Food Security

Food protection is consuming and being able to afford nutritious, healthy food — and plenty of that. That's a huge problem for many families around the world. Luckily urban farming is contributing to improved food security.

Creating a Sense of Belonging

Urban Farming is one way to bring together urban dwellers — to create a sense of community among otherwise autonomous and, in some cases, isolated citizens.

You get fresher, healthier food

Herbs, vegetables, and fruits — and are more likely to eat what's going on in the season when you eat what's being grown on an urban farm.

Provides a Learning Opportunity

Urban farms provide an opportunity for urban dwellers to produce their own food and learn in the process. They learn about different gardening techniques, the best fertilizer solutions, the sunlight needed, and temperature regulation, among other things.

Makes Effective Use of Land

We can use the land we already need to feed people effectively. Consider the rooftop or vertical gardens: they take up limited space but produce loads of safe, fresh food. Many hydroponic systems are set up vertically, including indoors.

Master a Special Talent

Farming ability According to the United States Farm Bureau, only 1 in 50 Americans has any background in farming (2 percent). That means that the average person on the road is much more likely to have other interesting skills such as speaking a foreign language (more than 15 percent of Americans are bilingual, and more than 50 percent of Europeans are). You are gaining a very specific ability by studying urban agriculture.

Conserve space

Urban farming allows extremely effective use of space. This is especially true in vertical agriculture. Vertical harvest, for example, one of the largest commercial vertical farms in the U.S., produces on 1/10 of an acre the same quantity of produce that conventional agriculture would need 5 acres to generate.

Highly safe source of food

The production of most urban farms is vegetables, usually loose-leaf lettuces, herbs, or brassicas. The health benefits of foods are almost unbeatable; according to resourse including 1) Vegetables do not have cholesterol. 2) Vegetables are rich in fiber. 3) Vegetables are sources of many nutrients including potassium, folate, vitamin A & vitamin C.

Helps avoid food insecurity

Food insecurity: an economic and social situation where access to sufficient food is limited or uncertain. According to the USDA, one out of seven Americans is suffering from food insecurity, with over 6 million children. Urban farming can help to ease some of the food insecurity pressures. For instance, urban farms developed in food deserts can be a source of nutritious food for nearby residents.

Easier to consume organic products

Organic foods contain fewer pesticides than traditional contain than traditional are fresher than conventional growing in your home in certain cases is the most sustainable way to grow in some urban farming settings you will need almost no pesticides or fertilizers, so they might not be fresher. Organic and non-organic production

No Need to Worry About Seasonality

If you grow indoors in a regulated setting, the erratic weather conditions don't limit your growing season. Climate conditions such as extreme cold or drought cost conventional farming trillions of dollars per year in lost yield.

Less Greenhouse Gas Emissions

"Some analyzes indicated that food-related greenhouse gas emissions had been minimized by taking agriculture to cities."

In other words, the same research connected above * does * stress that the savings in greenhouse gas emissions from urban farming are often overestimated, especially in high-density urban farming areas in the Northeastern United States (New York, Boston, etc.).

Lower than Purchasing Produce from Standard Supply Chain

On the basis of common sense, bringing your own fresh fruits and vegetables in time for growth would have a lower unit cost than shopping at the grocery store.

The explanation is simple: Items from the grocery store are heavily labeled. According to this Chron post, grocery stores mark up to 75 percent of the cost of their products, which is almost a 2x increase you pay. Beyond that, a lot of the original shipping expense of the produce comes from. The Outcome? It's way cheaper to raise yourself on average.

Increases Property Value

In a UC-Davis study on the benefits of urban agriculture, it was reported, "Studies connect urban farms and community gardens to increase home values and household income. The existence of gardens increased property values to as much as 9.4 percent within five years of the establishment."

Correlates with Socioeconomic Diversity

According to this report from the United Nations Food and Agriculture Organization, because of the conflicting social, economic, and ecological dimensions of urban farming (see figure), there is a connection between areas with urban farming projects and socioeconomic diversity. Dimensions of policies and forms of town farming.

"Farm Miles" declines (even in contrast with "local produce").

Food miles are described as the distance from where the food you eat is grown or developed. Local food miles created from an urban garden or farm can be less than 0.001 percent of the distance traveled from "farm to fork" by the grocery store (and even restaurant).

Think about it this way, and if you grow in your own house, we don't even talk about "miles" of food anymore, we're talking about "feet" of food!

Small Business Growth Driver

Urban and vertical farming are witnessing year-over-year growth in triple digits, and it is not slowing down early.

Less Infrastructure Investment

Needed While some vertical farming operations need tons of investment in infrastructure, the majority of urban farming production comes from small-scale CSA or small-scale farmers, according to this Ensia.com report, the amount of infrastructure costs needed for a greenhouse or indoor basic urban garden is lower than traditional farming.

Expanding Grant Funding Opportunities

In 2016, urban commercial farms provided more funding than any other year in United States history. If you are looking for grant funding as an urban farmer, at this pace, your chances of success can only increase in the future.

Improves state of mind

According to a Psychology Today report entitled "Plants Make You Feel Better," indoor or garden presence of plants:-Lower systolic blood pressure-Lower anxiety rates-Improve work satisfaction

Less Packaging Needed

If you harvest your food from an urban farm, you may be able to do away with packaging altogether.

Why is this a massive advantage?

Shipping is one of the most dangerous threats to the atmosphere on earth. According to this research on Livestrong, entitled "The Advantages and Drawbacks of Food Packaging," while packaging has advantages such as increasing shelf-life, there are also huge drawbacks: "Food packaging accounts for the greatest volume of plastic and paper waste, according to Duke University researchers Patrick Reaves and Michael Nolan, which makes up 20 percent of all landfills"

High food safety Outbreaks of salmonella or other pathogens on a wide scale are essentially a bi-product of the vast distance that food processing undergoes.

According to the World Health Organization (WHO), "Foodborne diseases are typically contagious or harmful in nature and are triggered by infected food or water bacteria, viruses, parasites or chemical substances entering the body." With the growing duration of the supply chain, the risk of contamination is increased.

Higher Food Quality

When going to the grocery store, you have little control over issues like -growing conditions- harvesting time-light exposure. It is a "take it or leave it" situation that you have very little control over.

You either buy a burnt carrot, or you don't get a carrot if you just want a carrot because all the carrots are heavily bruised.

Less Food Waste

And with urban farming, you can "harvest and consume," there is no disconnection between your supply of products and the amount you consume. Some consumer-level food waste happens when already-purchased produce goes wrong. If you just harvest what you're going to eat, you'll be wasting far less.

Correlates with Neighborhood Safely

According to a 2013 report by UC Davis (also stated above), 'Public gardens and urban farms build green spaces for recreation and enhancement of general concern for others in the neighborhood. 'm In essence, urban farms are basically the opposite of the Broken Window Theory, described as 'a Criminological theory of urban standardization and warning consequences farming.'

We need more town farms for all the broken windows in towns!

Aesthetically Appealing

The majority of humans find the sight of plants aesthetically pleasing in all cultures and backgrounds.

This data from Format magazine shows that plants are one of the most common artwork subjects of all time. No wonder so many urban farming companies take such impressive photos of their structures in high definition!

Capable of integrating into architecture

There are whole websites, such as Inhabitant, devoted to urban farming-themed farming. When you're an architect, you have countless possibilities and ways to integrate urban agriculture into your projects. Its architecture is one of the defining characteristics of any city, and all the incorporation possibilities are a big bonus for city planners and designers.

Wide Innovative Cities

Around The Globe, As you can see from this Pinterest Board on urban farming ventures worldwide, urban farming is gaining traction in areas around the world.

Areas such as:-

- Japan
- France
- Netherlands
- Kenya

What do you mean by that?

Wherever you are, you will possibly find other people in your town who are interested in urban agriculture!

Water conservation

Many growing types common in urban farming are much more productive with water than with general agriculture.

Hydroponic systems, for example, will use 2/3 less water than what would usually be required for the same amount of production, according to Lucky Roots.

Air Purification and Breathing

Based on data, entitled "The Five Benefits of House Plants," having some form of plant in your house can help with air purification.

This process has, in fact, been researched so extensively that we know that some forms of plants purify air better than others. Check this page for more information on different types of plants.

Could Increase Your Focus

According to a Scientific American research called "Houseplants Make You Smarter," broad data sets show plants improve concentration and attention span.

Rationale?

Human beings developed in environments with many more plants than most people are attached to in today's day and age than the computer screens. Although too much exposure to the screen is usually dangerous for attention span and concentration capability over, plant involvement results in the opposite effects.

Listen to kids, and if you haven't done too well on that last exam, you might need to do some urban agriculture!

Prevent Illness

According to Tree hugger's recent post, the presence of plants in your house or garden does play a role prevents the onset of sickness. It does not take into account the various studies it is known that actually eating more produce prevents illness.

Have you ever heard of Brassica?

This is one of the most popular vegetable forms by humans, including stuff like:-

- Broccoli
- Cauliflower
- Kale
- Cabbage
- Brussel Sprouts

Yeah, even mustard seeds come from Brassica, and there is no mustard without Brassica.

What is the point, then? If you try urban farming, you learn this type of knowledge, and as a result, you get a better understanding of what we eat.

Liberty

This is the greatest benefit: You can grow with urban agriculture:-what you want to grow-how you want to grow it where you want to grow in the end this is very strong freedom of speech.

1.2: Urban Farming Is the Future

As the human population increases, more people around the world tend to starve to death. Everywhere, people are hungry, whether its rural areas or urban cities. According to a U.S. survey Department of Agriculture, the number of people who are forced to live in areas with restricted access to grocery stores, supermarkets, or other sources of nutritious food has increased significantly. With the widespread understanding of global warming, shifting towards an eco-friendly living has become almost a requirement. The adoption of urban agriculture is one step you can take towards a sustainable existence.

Importance of urban farming in the future:

Urban agriculture, or generally referred to as urban agriculture, refers to growing plants and rearing animals that produce food within a town or region. It also involves processing and then the distribution of the output throughout the region.

Thanks to the technical upgrade, you can grow food in areas where it was difficult or almost impossible before. Urban farms may be traditional small outdoor community gardens or modern urban-designed vertical farms. Such futuristic farms can be built in various ways, but most have rows of racks lined with plants rooted in nutrient-rich soil, water.

The light

Plants expand using synthetically active PAR or photographic radiation. Since not all lights are ideal for plants, PAR reflects the amount of light that can assist with photosynthesis. PAR usually ranges from 400 to 700 nm of light. It is necessary to track the PAR to ensure that the plants obtain the required light. Now you can do smart agriculture without unpredictable weather.

Social aspects:

In addition, the Urban Agriculture & Forestry Resource Center (RUAF) has recognized the value of urban farming in areas affected by poverty and has collaborated with NGOs and experts to educate communities about the benefits of urban farming. With smart technology, urban farming can go everywhere, from simple community vegetable gardens to delivering nutritious food to customers in the surrounding area. Urban agriculture can be a blessing in urban towns and areas where ample space is a privilege. The urban farms at firmly place actually have plants of high density, which require very limited space.

Real-world example:

Across areas of Europe and Asia, there are some remarkable examples of urban farming. There is an urban farm at the Prinzessinnengarten in Kreuzberg, near the Berlin Wall, which grows a wide variety of vegetables and fruits in rice bags, recycled Tetra packs, and plastic crates. Another popular case in Singapore is Sky Greens, which is like a skyscraper on a vine. Considering that Singapore is one of the world's most heavily populated nations, it has little space for farming. Sky Greens has been tackling this problem by growing vegetables on a tall, narrow frame. The plants rotate gradually, so each crate gets enough sunlight exposure.

Plus, you can produce more in urban farms than just fruits and vegetables. For example, Urban Organics specializes in the growing of three kale varieties, two Swiss chard varieties, Italian parsley, and cilantro. Through a closed-loop method called aquaponics, it also uses the same water to lift the Atlantic salmon. In addition, fish waste is used to fertilize the plants that filter the water before it reaches the plants.

In reality, The Guardian has classified Urban Organics as one of the world's ten most creative urban agriculture ventures.

Future of Urban Farming:

It may seem at first that programs for urban farming in small communities will hardly have any meaningful effects. On the contrary, according to an Arizona State University study, if an urban farming initiative were introduced in each global city, the urban agricultural industry could produce up to 180 million metric tons of food annually. This is a huge number because it is roughly equal to the agricultural production of 10 percent of the earth. Additionally, urban farming also helps with financial savings from decreased storm water runoff, urban heat-insula effect, pest control, energy costs, and could theoretically add to $160 billion annually.

It makes sense, with so many positive outcomes, why researchers and environmentalists promote urban farming and make people aware of the big global, cultural, social, and environmental issues.

The planet is rising more food than ever, and yet the world continues to starve millions of people. Everywhere people are hungry-in the region, in the suburbs.

But increasingly, in the war on hunger, one of the front lines is in cities.

As urban populations expand, more people find themselves in food deserts, places with "restricted access to supermarkets, supercenters, grocery stores, or other safe, accessible food sources.

New technologies change the equation, enabling people to grow food in areas where it was difficult or impossible before, and in quantities comparable to conventional farms. Urban farms can be as simple as Common outdoor gardens in small groups, or as complex as vertical indoor farms where farmers think in three-dimensional terms about expanding space.

Such large, futuristic farms can be built in a variety of ways, but most of them contain rows of racks lined with soil-rooted plants, water-enriched by nutrients, or only air.

The growing tier is equipped with U.V. light to replicate the sun's influence. In contrast to the unpredictable weather of outdoor farming, indoor growing allows farmers to customize conditions to optimize production. Farming can go anywhere, with the proper technology. That's what the latest trend in urban farming shows — these farms go beyond mere community vegetable gardens to provide food to customers in surrounding areas. All the vertical farmers need is some space and electricity access; no special facilities required. Farmers can buy whatever they need to start and keep their farms.

In reality, since access to starting materials is so simple, officials don't really know how many urban farms operate in the U.S. A 2013 survey conducted by the National Center for Appropriate Technology (NCAT) received 315 responses from the operators who identify them as urban or suburban farms.

Federal agricultural development grants, however, indicate thousands of city-dwelling applicants, suggesting that the number of urban farms is likely to be substantially greater. "Unlike square foot, you have to look at certain services in cubic feet.

We can actually bring out a lot of produce from a facility like this, "told Futurism Dave Haider, the president of Urban Organics. Technology enables vertical farmers to monitor their farms' climate, enabling them to grow far more in the same amount of space.

Urban farms are more likely to expand than just fruit and vegetables. Urban Organics grows three kale varieties, two Swiss chard varieties, Italian parsley, and cilantro, but uses the same water to collect Arctic char and Atlantic salmon — a closed-loop method sometimes referred to as aquaponics. Fish waste fertilizes the plants that clean and filter the water before returning to the plants; excess drops into the fish tanks. In 2014 Urban Organics opened its first farm inside an old brewery site.

Throughout the years since food has been delivered where it's most needed: to people in the Twin Cities food deserts.

In 2014, The Guardian called the company one of the world's ten most creative urban farming ventures. "It is also one of the driving Strengths behind our first farm which actually was situated in a food desert, to try to make a dent in the industry when it comes to food deserts," Haider said.

Urban Organics markets its produce to local supermarkets and supplies fresh fish to restaurants in the area. "That was sort of a kind of our approach — let's try to develop high-quality protein in an environment that needs it the most." As more people migrate to towns, problems like food shortages may get even worse. The vertical farm is eco-friendly too. Aquaponics processes result in scarce waste. Vertical farming makes it possible for farmers to use their limited area more effectively, and we can collectively make more use of existing space rather than creating more arable land, leaving more habitats intact. By putting farms near to vendors and customers, fresher goods will hit tables with less truck dependency, which leads to emissions and global warming.

Farms of the future:

When people continue to research and tweak urban farming methods, we can begin to learn more about how the environments around them and the broader global community will benefit. Data on how urban farms impact their local communities directly could push lawmakers to support and invest more in urban farms.

To this end, Gordon-Smith has planned another side project: a whole building or community to check the technologies for urban farming while collecting data. While the venue has not yet been Chairman of Brooklyn borough Eric L.

Adams vowed has already given Gordon-Smith a $2 million commitment, he has already submitted his plan to the New York City Council. The plan is pending approval from the Land Use Committee, and no indication is given as to when it will be determined. Vertical agriculture, and urban agriculture, in general, maybe a significant boost for areas with the capital to invest, feed people and support the local economy. Nevertheless, it is important to realize that urban farming is not a special solution to a global problem, such as helping people access enough nutritious food. Gauthier, the urban farming expert at Princeton, points out that there are many essential crops that simply cannot be grown indoors, at least nonetheless. "We shall possibly never go to produce soybeans, wheat, or indoor maize," he said. "Vertical farming isn't a worldwide solution to hunger. It's not the solution, but it's definitely part of the solution. "Other attempts to combat world hunger are giving people in developing nations more economic independence by providing them credit lines or putting in place basic income schemes, such as those implemented in Kenya. Education, social change, and empowerment of women are all social programs that will help more people get access to the food they need. Supporting one another, including their families. Breeding farms have the ability to alter the agrarian landscape around the world. Granted, we certainly won't see a world with super cities where all the farming takes place in high-rise buildings. Yet urban farms can deliver higher yields in smaller areas, increase access to healthier choices in urban food deserts, and reduce the effect of feeding the planet on the environment. That seems to be enough of a justification for these innovative farming practices to continue to grow and expand.

1.3 Difference between Urban and Rural Gardening

Human settlements are categorized as rural or urban, based on the density of structures created by humans and inhabitants in a given location. Urban areas may include cities and towns, while rural areas may include villages and hamlets. Although rural areas will grow spontaneously based on a region's available natural vegetation and fauna, urban settlements are suitable, planned settlements built up according to a process called urbanization. Rural areas are mostly concentrated on by governments and development agencies and converted into urban areas. In comparison to rural areas, urban settlements are characterized by their advanced public services, educational opportunities, transport facilities, business and social activity, and a better overall standard of living. Typically the socio-cultural figures are based on an urban population.

While rural settlements are more focused on natural resources and occurrences, the urban population receives the benefits of the advancements made by man in the areas of science and technology and is not dependent on nature for their day-to-day functions.

Businesses remain open in urban areas late into the evenings, while sunset in rural areas means that the day is almost over. The flip side of this is that there are no emissions or traffic issues in rural areas that beset typical urban areas. While several governments concentrate on rural development, they have also attempted to 'secure' these areas as protection of the basic culture and traditions of their country.

Residential areas are also classified according to land use and population density. But this can vary from developed to developing countries. In Australia, for example, urban areas must have at least 1,000 inhabitants with 200 or more people per square kilometer. In China, the density criterion for an urban area is around 1,500 people per square kilometer.

Human settlement is generally divided into two categories, i.e., based on population density, growth, services, job opportunities, education, etc. Urban and Country. The city refers to a human society where there is a high rate of urbanization and industrialization. On the other hand, it is one in a rural settlement where the urbanization rate is very sluggish. The important distinction between the two human societies is that while urban areas are densely populated, rural areas have comparatively fewer populations than urban ones.

Definition of Urban:

The term urban refers simply to the densely populated region or area which possesses the characteristics of the man-made environment. The citizens who live in this region are engaged in industry, or services. There is high-scale industrialization in this settlement, which results in better job opportunities. The urban settlement is not limited to the cities alone, but it also includes towns and suburbs (suburban areas).

There are many advantages of living in urban areas, such as convenient access to different services, better transit facilities, options for entertainment and schooling, health facilities. Since it suffers some disadvantages such as pollution caused by large-scale industrialization and transportation means such as buses, trains, cars and so on, leading to an increase in health problems among the people living in that city.

Definition of Rural:

We describe the term 'rural' as an outlying area. It refers to a small settlement outside the boundaries of a town, commercial or industrial area. These can include areas of woodland, villages, or hamlets, where natural vegetation and open spaces occur.

The population density in this area is small. The residents' primary source of income is agriculture and livestock husbandry.

Cottage Industries form a big source of income here as well. In India, as per the planning commission, a town whose population is below 15,000 is known as rural. Gram Panchayat is in charge of looking after these regions. Furthermore, there is no municipal council in the villages, and there is a large percentage of the male population engaged in farming and related activities.

The key difference between urban and rural:

The fundamental following points deal with the disparities between the urban and rural areas:

- Urban is defined as a community where the population is very large and has the characteristics of a built environment (an environment that provides basic human activity facilities). Rural is the geographical region in the outer parts of the towns or cities.

- Life is quick and complicated in urban areas, while rural life is simple and comfortable.

- Cities and towns are included in the urban settlement. On the other side, there are villages and hamlets within the rural settlement.

- In urban areas, despite the presence of the built environment, there is greater separation from nature. In comparison, rural areas are in direct contact with nature, as they are influenced by natural elements.

- Urban people are active in non-agricultural jobs, i.e., in the manufacturing, manufacturing, or service industries. In comparison, rural people are mainly occupied by agriculture and animal husbandry.

- Population wise, heavily populated metropolitan areas, which are focused on urbanization, i.e., the higher the urbanization, the higher the population. The rural population, on the opposite, is small and has an inverse relationship to agriculture.

- Urban areas are built according to the process of urbanization and industrialization in a planned and systematic way. Rural production is limited, based on the region's abundance of natural vegetation and fauna.

- Urban people are highly active when it comes to social mobilization because they often change their profession or residence in search of better opportunities. However, there is comparatively less extensive industrial or territorial mobility of the people in rural areas.

- At the time of job distribution, the division of labor and specialization is still present in the urban settlement. There is no division of labor, as opposed to rural areas.

Is Urban Gardening More Important Then Rural?

Urban agriculture or over the past few years. Growing food in urban areas has become It has become trendy; it is believed to be food's future, and new "smart gardening" brands are growing faster than ever before. But what makes urban farming in the century of urbanization so attractive and apparently superior to conventional farming?

It is more competitive

With the aid of urban agriculture, one can produce as much as 100 times more food than with normal (per square foot) farming. How could that be? It's all about the path-most urban farms that are vertically built, which allows producing on a square food to expand as many levels as possible.

So, if you have 2 square feet free room in your flat, instead of having a corner of lettuce soil growing there and molding your beautifully finished walls, you can actually have a tasteful-looking urban farming system of lettuce growing in the shelves or behind the walls.

Have you ever heard of weather raising yield in a given year? In fact, the weather is a very, very minor factor for food production by urban farming, as it typically takes place indoors and relies on the built water network, artificial lighting, and nutrients already worked-in in the soil.

It's more sustainable:

Most urban gardening systems result in significant savings in water, power, and energy. When we talk about the urban farming systems Click & Expand, they use about 90 percent less water and four times less room compared to conventional farming. Many point out that starting an urban May be costly to plant. Currently, $500 could be expensive. In fact, a $500 urban farm would "earn back" itself in one year. But why do they make the investment? The thing is that the savings in energy, electricity, and space are not all financially sustainable. The more we move our daily food production to urban farms instead of traditional farms, the more we decrease drought growth, soil erosion, and similar problems.

Organic farming becomes more accessible:

Urban farming encourages farmers to grow crops in an even more regulated and aware way, leading to more opportunities to grow organic food without additional investment.

How could that be? The key reason anything is not organic is that when the environmental conditions are not crop-friendly, the farmer is forced to use chemicals. As stated above, the environmental factor in urban farming is reduced to a minimum by the weather, so there is no real need to use regulators for chemical production.

It's a small space friendly:

Yeah, you can have here and there a typical herb pot, but it won't feed you. And if you want to grow enough to use the harvest straight for at least a week, you'll be forced to have flowerpots everywhere. Everywhere, literally-on the sink in your kitchen, under the sink in your toilet, under your bed, around your bed, on the shelves, in the cupboards.. Oh wait, they've got to need energy. Bigass windowsills everywhere around the house. Completely unlikely, is it? Urban farms make life much, much simpler, in the area. Most are small, neatly built "closets" with several shelves, capable of growing tens of different plants on each shelf. Sounds a bit better than lying on your bed with a bowl of basil, doesn't it? It is not only a food-growing unit but also an aspect of design, a piece of argument. And you can even put it in the darkest possible corner.

It allows you to enjoy all-year-round fresh produce

Asparagus grows in spring, basil has to be sown from March to June anywhere, and strawberries can be harvested in late June and in July. If we want to earlier have a certain crop, plant engineers and pesticide producers need to put their heads together to find out something. Since the environmental and seasonal element in urban agriculture is reduced to near zero, everything can be harvested at any time. You can overlook the best seeding times indicated on growing plant's seed packets and climatic requirements. Would you like fresh, homemade strawberries in December, or freshly grown lettuce every two weeks? No issue whatsoever. When one plant has grown and is ready for harvest, it can be replaced with a new soil & seed cartridge immediately. No need to, no hassle.

It's easy

If you haven't already realized it, it's really easy urban farming. In urban farming, the gardening skills and green thumbs required in traditional farming are being replaced by technology.

If you own an urban farm, you can grow fresh food for yourself and your family all year round without ever having to touch the soil, get dirty, put seeds in any location, or fertilize.

It has already been taken care of, at least with Click & Expand systems. Everything you need from your side is to take off stickers, stick cartridges in the right places, and press buttons in an app to water the plants. If the farm is asking you to.

Chapter 2: Planting a Garden in an Urban Area

Working in an urban environment doesn't mean you have to quit gardening. Building a garden in what little room you have taken only a little preparation and planning. You will grow them if you find a place for your plants. When it comes to growing plants, urban dwellers are limited. Depending on where your home or flat is situated, you may be wondering where to put your plants. This is a good spot if you have a window or loft area.

Even if you don't have much sunshine, plants can be found, which will grow. The rooftop access will give you, even more planting choices. Also, be sure to take wind into account, so the safety provided by the fencing panels is strong.

Wind may either dry the leaves out or damage the stem. Use fencing to create a brisk breeze.

2.1 Making a Garden in Urban Area

Living in an urban areas, it doesn't mean you have to quit gardening. Creating a garden in what little room you have taken only a little information and preparation. You will grow them if you can find a place for your plants.

Designate a location:

When it comes to growing plants, urban dwellers are limited. Depending on where your home or flat is situated, you might be wondering where to put your plants. This is a good spot if you have a window or loft room. Even if you don't have much sunshine, plants can be found, which will grow. The rooftop access will give you even more, planting choices. Just be sure to take wind into account, so the safety provided by the fencing panels is strong. Wind may either dry the leaves out or damage the stem. Using fencing to create a brisk breeze.

If you have trellis, make sure it is secured correctly so it won't float away from an unforeseen wind rabble.

Pick the right containers:

Rooftop gardens are ideal for urban surroundings. When you are gardening in an urban environment, you will need plenty of containers for planting. Choose the containers that are wide and deep enough to grow and thrive on your plants. Ensure sure the pots have a drainage hole to the rim. It is also necessary to weigh the containers.

If the weather changes, don't buy anything that's too big to push. You may do want your containers to fit with the rest of your home decor. Self-watering pots are available if you are concerned that your garden could overwater or underwater.

Consider the soil:

The soil also isn't suitable for planting in an urban environment. It also contains too many rocks and inadequate nutrients. Invest in a premium potting mix. Fertilization is essential, as well. During the growing season, a fish emulsifier can ensure that your garden is fed properly. Compost is another organic fertilizer ideal for your urban garden too.

Watering plan:

All plants must get an appropriate quantity of water. Not enough water will cause leaves to dry or drop blossom. But too much water causes its own problems, including root rot and infestations of insects.

As mentioned earlier, the best option is self-watering containers. The attached reservoir is simply filled in, and the plants can absorb moisture as needed. If you can rely on irrigation, make sure that the soil remains moist, without being soaked. You must create an area where your pots can drain easily if need be.

Choose your plant wisely:

While you may be tempted to expand anything, there's not likely enough room for you. Steer clear of seeds, such as pumpkins, which appear to overrun gardens. They don't get along with other plants well and can compete for space.

You need to ensure that the plants you select grow in accordance with the crops that surround them. Root vegetables, like carrots, are not the perfect alternative, either. In a potting area, they won't have a lot of space to grow, which can hinder their growth. Herbs can grow pretty much everywhere, even in small containers. Tomatoes, peppers, and beans are rising rapidly and producing an impressive yield.

2.2 Steps of Planting a Garden

When winter recedes, and we dig out our slightly lighter clothes from the back of our closets, our minds turn to vegetables and flowers — even if our expanding spaces are confined to balconies, rooftops, fire breaks, sunny windowsills and small backyards for city dwellers. Yet we should not be deterred by these small spaces; it is surprisingly easy to grow all sorts of beautiful and delicious plants even without a lush suburban yard.

Steal my sunshine:

When your growing room has only a few hours of sun a day that could require anything like a sunny window, there are still choices for you. You're going to need a jar, first. You may be shocked that, while beautiful, these costly glazed or terracotta pots give absolutely no benefit to growing plants besides looking nicer than, say, a plastic bucket. "For herbs, plastic quarter take-out containers will work great, or you might get an expensive glazed pot, and that works fine too.

The objective is to ensure that the container has a hole in it to provide adequate drainage, preferably along the edges, around one inch from the bottom upwards. You are a noob in the garden and can unintentionally overwater, so drain holes allow any excess water to escape without the roots rotting. You might insert a small piece of glass or a bit of cloth to prevent soil from falling out of the bottom, but you'll always want to place a jar like this on a catch saucer in the event of excess water falling out. Hopefully an old T-shirt would fit as well. (Typically I've only done without and allowed a little soil coming out of the bottom with excess water.) Be vigilant with window boxes hanging from a window ledge: for one thing, they can be unsafe if not properly mounted, and for another, it's more difficult to plant a bunch of different plants in a single container. You will compete for water or nutrients with each other, and if they have different water needs, then you could end up over- or under watering something. You would also need soil; soil can be toxic in most urban areas, and usually rocky, sandy and nutrient-depleting. Better to buy some potting soil. To beginners, Miracle-Grow or other boosted-nutrient soils work great, but O'Brien suggests going organic to advanced gardeners. "If you're using Miracle-Grow or stuff like that, it's a major nitrogen shock all at once, so it doesn't last in the soil," she says. Nitrogen can soak into groundwater, which is not healthy, but it also has odd effects on the plant.

O'Brien says she's found that too much nitrogen can lead to more succulent leaves, which can attract pests and fewer fruits. You're going to want to get some fertilizer too. Just what kind of fertilizer you're supposed to use varies by plant, soil type, and pH levels, but to get started. So what can you develop at a rough, shady spot? Edibles can be tricky but without a lot of sun a few herbs can grow happily. As with chives, mint is a great choice. On your side, they grow without much effort, and are heartfelt enough to withstand your errors.

Beginners may want to go with seedlings rather than seeds. And that is definitely not what you need. And some plants are difficult to grow from seed: "Parsley, like, can germinate for three weeks," she says. Don't bother: Take some seedlings from a local garden shop or market for farmers.

So, you have a little more sunshine than just a window — may be a deck or a window facing south. Now you have choices. First of all, go bigger: Your container size will greatly affect your final plant size. Always go bigger and never overcrowded plants than you think you like. Those giant Home Depot five-gallon containers are perfect for a single tomato plant or chili pepper plant, believe it or not. Don't attempt to bring in more; instead of growing tall and solid, the plants will battle each other.

You can make a pretty spectacular drainage system out of those buckets for intermediate gardeners. Fill them with gravel up to a few inches from the bucket's rim, then place a layer of cloth (weed blocker is good but honestly cotton will work fine) over it, then fill the bucket with soil the rest of the way. This technique gives you a much more controlled drainage system and helps avoid the mold and sitting water in your soil. Be cautious when planting several plants together; take a look at the tags to ensure they have the same sunlight and water requirements. Obviously, some plants function together; "If it was a box of herbs and you had parsley, basil, mint and cilantro? They're all kind of close, they like a lot of water, and they'd do their best under the full sun, "O'Brien says. Yet some require less water and less heat, such as garlic, thyme, and oregano, and wouldn't do as well under the same conditions. Environments with part sun really open up what you can expand. Greens are extremely simple and a perfect way to play around with seeds are for beginner gardeners. Lettuces, arugula, and chard are all super easy to grow, and they germinate quickly.

In the end, beginners and intermediates alike may also opt for certain non-leaf edibles. Tomatoes are very easy to grow, but be careful what kind of tomatoes you choose: O'Brien prefers cherry or grape tomatoes that produce harvests all summer long and can handle less sunshine and less space much more readily than, say, a beefsteak. Even chili peppers are extremely easy to produce. If you start with the seed, from the seedlings? Only move over to the market or hardware store for your farmers.

You'll definitely also find pests with part heat. There are so much bugs (and even squirrels and pigeons) even in an urban setting that are just as entertaining as you are to chomp down on some new local produce. It doesn't have to be difficult to fix bugs: a quick solution of a tablespoon or two dish soap to a quarter of water in a spray bottle and squirting on leaves and stems should deter most bugs.

As for irrigation, you can go as easy or as complicated as you want again. The only real rule is not to over- or submerged and flush the stem directly into the soil. Although some plants can enjoy a nice misting of water on their leaves, don't ever water the leaves-they can get moldy and die. (Refer to the seed packet or add once again.)

Black whole sun:

You have rooftop access or a spacious backyard, so now the sky's the limit — or rather, the limit is your available space. A main mistake made by many early gardeners is to pick the wrong crop type for a limited area. Root vegetables are fun — "To pull a carrot out of the ground is really something, but since you're eating the plant's root, you only get one harvest each year, which doesn't make it a good use of the room. Ditto to giant pumpkin, butternut squash, and watermelon seeds. They are going to grow, but not all that well; these plants seem to want to cover the whole soil, and do not play nice with other crops. And forget about corn: It's very fun like carrots, but the amount of room you need to grow even a meal worth of corn is probably more than you've got in total.

Use smaller fruits and vegetables instead: tomatoes and chili peppers, yes, but also zucchini, aubergines, cucumbers and beans. Of these, follow the instructions on the package. Others are climbers, like the cucumbers and beans using an extremely old Native American technique, intermediate urban farmers may plant many items at once, stake or trellis your climbers, then grow underground plants such as greens or herbs beneath them. For the same amount of space you receive twice the crop. For around $18.00, you can get a cheap wooden trellis, or you can go for stunning expensive wrought-iron trellises like this one ($387, which, obviously, would be crazy), but they're going to work the same way.

For full heat, instead of using pots, you may want to study an elevated planter. This is a little bit more work, but it may be worth it: You give your plants a little more space to stretch their roots and develop bigger and stronger "square foot gardening" system, which blocks off an elevated bed into separate sectors to ensure that other plants develop unimpeded in a raised bed. And you can buy already put together the boxes or other setups for that device or you can cut for a simpler wooden one. If you have a more complex raised bed greenhouse, finding out some form of automatic watering device is worth it. That will allow you to leave for a day or a week and not worry about wasting all your hard work, and will make sure that your plants are happy.

The best gardening gadgets:

As for automation, off-the-shelf solutions are not lacking to help you cultivate herbs and flowers. Some even are pretty sweet! But they certainly aren't as inexpensive as a DIY solution, and many of them won't give you the eternity to learn the basics of gardening which will allow you to advance to trickier plant varieties. Now, any way you get into the joys of having new greens at hand is a win.

Gardening gadgets can be loosely divided into two categories: tracking devices that control soil dryness or nutrient balance, and all-in-one growing systems that usually forgo soil for water-based, hydroponic gardening.

Of all the monitoring gadgets — the Parrot Flower Control, the Oso Plant Link, the Spruce Irrigation Network, there are only a few that I would personally recommend to the casual gardener of the apartment, the Chirp! Ad fruit water sensor, which is a basic stick that sits in a pot and creates a bit of buzzing noise when things get too dry; or Dr. Meter hydrometer, a clever little analog tool that doesn't even require batteries to operate. Both are highly rated (and considered) and do the only thing you really want, which is to give you a second advice about when watering is appropriate. And because you are likely to water all of your plants at once, using one display in a single pot would be enough reminiscent that your plants need a little snack. (Bigger pots and larger plants can pass through water at different levels, but you can still simply shift the monitors to double-check what your finger tells you in the soil.) Another basic instrument? Reminding you to give your plants a drink every two or three days.

If you want to bypass the pots and soil entirely, there are many all-in-one hydroponic kits around these days, including the groundbreaking category Aero Grow / Aero Garden, depending on the add-ons. Such modules are just as basic as they come. A tub of water at the bottom, some hanging, foam-filled seed pods in the center, and some kind of rising light at the top. Many used the first Aero grow nearly a decade ago and found it very enjoyable, albeit without the pleasure of playing with the water. Such kinds of systems come with their own drawbacks, you can't really grow root vegetables and you're going to have to cut them down every month or two to clean up the insides and prevent them from being scummy with algae. That said, they couldn't be easier to use and while the prepackaged seed pods are hilariously costly compared to buying seeds in packets, I ripped out the old root systems and planted new seeds without any problems. There are a few new competitors in the market, such as the Click & Grow, which has an ever-so-slightly more classic design, or the hydroponic Modern Sprout device, which is clad in real wood and attractive. They all work exactly in the same way, so if you're only trying to get bunches of fresh herbs on your shelf — and the units with rising lights make it possible, say, even in unwindowed kitchens — you can't really make a bad choice.

There are a few other types of gardening devices that can be safely overlooked for now, super "smart" planters with integration of Internet-Of-Things and other whiz-bang apps, such as the AliGro or the Planty nothing pot, which are interesting but much too pricey and unproven for the beginning gardener; and grow lights. If you feel absolutely that rising lights are a cure for your black thumb — and as the owner of a few in my apartment garden, I can assure you they aren't — look at some of the newer, LED-based variants that use a little less energy than incandescent models. Nevertheless, be ready for an apartment full of purplish-blue sun, and for neighbors who knock at your door asking about your stash.

Fine, I'll grow something:

The real lesson here is that it's incredibly simple, relaxing and satisfying to garden. A take-out container that used to carry wonton soup, plus some soil and a good little seedling from your farmers' market, will yield spices, berries, vegetables or flowers. Don't overthink it /You're going to make mistakes, You're going to give up on few plants to inexpertness, but growing plants is a way to improve your living space and help you quiet — if it's only a few minutes a day — the insane bustle of urban life.

2.3 Why is It Difficult to have a Garden in Urban Area?

Growing produce can be an amazing experience in your own backyard or in a community garden that allows you to pick not only the things you eat but also to track the seed-to-harvest process. Issues surrounding urban gardens usually aren't before your mind when you decide it's time to open the ground in your yard or rent a garden plot, but there's a lot more to think than just buying your seeds.

Problems with an urban garden

Some problems with urban gardening aren't readily obvious when you first dig the soil, but they are real. Here are some of the common things you should remember before planting:

Permits:

You can need a permit to dig up the lawn, build a fence or hold urban livestock, including chickens, bees, and goats, depending on where your garden is located. Before you put your dreams in the garden, consult with your local municipality to avoid finding out the hard way it is not permitted. A lot of problems with urban gardening can be avoided by first procuring the correct permits.

The human element:

We always like to believe that our neighbors are both helpful and tolerant of our garden efforts, but that isn't the reality always. Until beginning a front yard garden and erecting a fence where there is a lot of foot traffic, it is a good idea to talk to neighbors. Produce theft is a real problem and is happening everywhere to frustrated urban gardeners.

Contaminated soil:

Even if the soil is good and rich in your urban garden, it can conceal hidden pollution from past times. Lead exposure is by far the biggest danger, and while most vegetable plants won't take the advantage into their system, it can be a concern if you don't wash the produce properly or if a child eats the garden soil. Before you get to the planting, a soil check for heavy metals is good practice.

Ozone:

Burning gasoline and other fossil fuels can cause near-field ozone pollution. While you can do nothing to protect plants from this threat, understanding that ozone is a concern will help guide your gardening efforts. The growth of ozone-resistant garden plants, but are not yet available to the public. Before then, you may want to move the gardens further away from roads and pollution sources into areas.

Water supply:

Rainwater gardening is romantic and earthy, but not every area has rainwater that is safe for gardening purposes. Within urban areas, contaminants can accumulate within rainwater, damage plants, and cause possible harm to gardeners. Based on natural minerals and contaminants, such as fluoride, that can damage sensitive plants, and municipal water may also be suspect. For some areas obtaining usable water can be a trick, particularly where drought and water rationing are popular. Prepare your water ahead long before you start planting.

2.4 Planting Flowers and Vegetables Together

A garden planted with each type of separated flora or vegetable can look well organized, but separating plants lead to dependence on herbicides and insecticides. Instead, you can use the seeds to plant flowers and vegetables together to attract beneficial insects and birds to help keep crop pests under control. This is not about planting a couple of flowers in your vegetable garden. Many vegetable plants have beautiful flowers and foliage, as well as colorful vegetables that add value to your flowerbeds when planted. When you fill in the empty spaces between plants, you'll will the area for rising weeds.

- When selecting companion plants, fit the flowers and vegetables according to their sunlight and water needs. Vegetables need full sun to grow, but that doesn't automatically mean you can only plant full sunflowers. Tall plants, including tomatoes, can give shade to flowers that grow in partial sunshine or partial shade.

- Plant maize in clusters of at least four short rows, rather than one long straight line. The morning glories of plants (Ipomoea spp.), nasturtiums (Tropaeolum majus), or other flowering vines between the corn stalks and train the vines to expand the stalls.

- Plant a cabbage and lettuce course, selecting a combination of green-leaf varieties and plants with red, yellow, or variegated leaves. Plant cabbage or lettuce plants within six inches of each other. Fill annual flowers in the spaces between vegetable plants which suit the colors in the leaves of vegetables. Petunias and pinks come in pink and purple shades that mimic the hue of green leaves. You can also plant some white flowers that accentuate the white colors of some vegetable leaves and serve as a backdrop to highlight the orange, violet and pink leaves.

- Instead of other large-leaf plants, such as the elephant ear, grow kale in your ornamental beds. The broad leaves bring to the garden a solid structure but the ruffled leaves bring a delicacy that works well among flowers.

- Surround cucurbit crops, including cantaloupe, watermelon, cucumbers, pumpkins and squash, featuring a wide variety of annual flowers to draw insects into the planting area to ensure pollination and fruit output. Unless the flowers are not pollinated, certain plants do not bear fruit. Pepper plants can grow peppers without pollination but the yield is greatly increased by pollination.

- Snap peas in flowerbeds in place of flowering ornamental vines. Snap peas feature delicate flower flowers, elegant leaves and twining vines, with the added benefit of edible pods being made.

- Plant a variety of marigolds (Tagetes spp.) in any vegetable garden in order to discourage unwanted insects which may kill crops. You can cultivate perennial marigolds in frost-free zones around your vegetables for permanent pest control. Perennial marigolds are popular canyon marigold (T. campanulata), Mayan marigold (T. nelsonii), and Mexican marigold (T. lemmonii). Colors involve various orange and yellow colors, although other flowers may be variegated with different shades of the same hue.

- To eliminate weeds, add a 3- to 4-inch layer of shredded bark mulch around all flowers and vegetables in the yard. Planting flowers close to each other between vegetables greatly reduces the amount of weeds, but between plants there are still empty spaces where weeds can grow. Should not explicitly move the mulch against the stems of the plant.

Tip

Just as you often have to harvest vegetables, such as squash and indeterminate tomatoes, to keep them bearing fruit, you often have to cut flowers or expired deadhead blossoms to extend the time of bloom. Crop development ceases and the plant grows seeds without cutting or deadheading.

There is no rule that can says you can't combine vegetables and flowers. The vegetable garden actually greatly profits from introducing some flowers and herbs. Yet it is not just beauty that will allow flowering plants into the vegetable garden. Companion planting garden flowers and herbs provides many beneficial features that can protect your crops from insect pests, and make them even more successful.

If you can't repel an insect, throw it into a plant for sacrifice. Sometimes this is achieved with another vegetable crop, like surrounding cabbage with a trap crop (or capture crop) of collards to attract the diamondback moth. The pest insect must collect on the trap crop which is eventually pulled and disposed of. Probably the most common flower trap crop is nasturtiums which attract aphids. Nicotiana is nice to that as well. Chervil protects the leafy greens away from slugs, and mustard draws Lygus bugs (tarnish bugs) away from apples and strawberries.

Weigh the possibility of bringing more of the pest into your garden than before, before you plant trap crops. The method is usually used the year after the plants have sustained serious damage by an insect. Time it so that, if you can, the trap crop is a little more mature than the plant it protects.

Vegetables do not always have spectacular flowers. To ensure that the bees find your vegetable plants, companion plant flowers with high concentrations of nectar, or in shades of blue, yellow, or white. Don't forget the plants in bloom. Herbs in the mint family, such as thyme and oregano, are bees' special favorites. You'll need to avoid harvesting a few plants, of course, to give them time to grow buds and bloom. Other options include cosmos, larkspur, mints (watching for invasive or placing in a container), sunflowers, sweet peas, and zinnias.

Not all species are horticultural pests. Some insects actually feed on the pests. Insects, including ladybugs, lacewings, parasitic wasps, and ground beetles, are included. Like like every other species, plants have some preferences for beneficial insects. Companion plants their favorites, and soon you will have good insects searching for the bad insects. Peregrine, dill, cilantro, and aster family flowers are especially good for attracting beneficial insects.

In short, the idea of biodiversity is what all this companion planting is leading to, or planting a large range of items rather than a single monocrop. This helps confuse insect pests by planting things that they enjoy with things that they won't like as well as attracting beneficial insects that can keep pests under check. Whether there is also a common relationship between different species of plants is still under study.

The last benefit of a partner planting flowers in the vegetable garden is the freedom to position your cutting garden where its design or appearance will not be judged on. If you want to plant black-eyed Susan, salvia, celosia, and zinnias in straight rows that are often cut halfway down, plant them with the vegetables, where looks don't matter as much as work. As cut flowers and pollinator lures, let them do dual duty. Companion growing vegetables, herbs, and flowers is how the original form of cottage garden developed. Sectioning off gardens for specific plant types was a privilege for the wealthy. Besides all the benefits described above, if you're short of space or time, the solution to your gardening problem may be companion planting.

Chapter 3: Container Gardening in Urban Areas

Container gardening is when plants are cultivated rather than planted in containers such as pots. Container gardening is for urban areas where it is not practical to actually have a greenhouse the benefit of container gardening is that you can reuse old containers for your garden around your home, so it's budget-friendly and environmentally friendly. Container gardening is nice too because urban birds profit from it.

3.1 What is Urban Container Gardening?

Container gardening is when plants are cultivated rather than rooted in containers such as pots. Container gardening is for urban areas where it is not practical to actually have a greenhouse. It is space-efficient and versatile, so it can be designed to suit your garden wherever you want to. The benefit of container gardening is you can reuse old containers around your home for your garden so it's budget-friendly and environmentally friendly. Container gardening is also smart, since urban birds profit from it. Birds living in urban areas took benefit greatly from plants they can feed on, such as berries, and by planting some of these at home, this helps to increase the bird population.

What kind of plants can you grow on the container?

There are several plant forms that can be grown in urban areas. Some grow different berries and flowers. Below is a list of common berry plants that attract and feed birds. For guidance on how to plant them? Follow the list:

Berries:

- **Blueberries**

- Chokecherry

- Black chokeberry

- **Winterberry**

The flowering plants are several other types of plants that grow well in containers. Some flowering plants attract birds, and they serve as shelters. They are also stunning, and will accentuate any area you want to transform into your urban garden! In addition, flowers provide birds with seeds and nectar.

How to plant blueberries?

Blueberry grows very well in containers; during the spring, they produce beautiful white flowers, and at the end of summer, they produce delicious dark blueberries. Blueberries are simply delightful; you can make your whole family happy with them while feeding the birds in your neighborhood.

Depending on the plant type, blueberries grow in shrubs that grow up to 5 and 10 feet in height. For this reason, make sure that it is of the smallest variety for the environment of your zone when planting blueberries in containers. If necessary, get the container also hold of acidic soil as blueberries grow best with some acidity in their roots.

Instructions:

Plant with good drainage in a wide container and ensure the plant gets enough sunlight. You can use any kind of soil to plant in a container, but if you can mix in around one-third of acidic soil, that will work better.

Keep the soil moist, but not hot enough to create mud or pool on the surface. A lot of colds.

If you have enough room plant in containers for more blueberry plants, this will help with pollination that will make your plants yield better berries and a larger amount. Be patient, and it can take a few years for your plant to grow significant quantities of berries. See to it that your blueberries are covered in winter. Place them inside a building or cover them with burlap to keep the plant from dying from the cold weather.

Often plants obtain minimal nutrients in containers, use some compost fertilizer within the container to help your plant be healthier and grow more fruits.

How to plant chokecherries:

The Chokecherry plants are well-growing shrubs in large containers and are easy to maintain. Its aromatic white flowers form beautiful cylindrical bouquets, measuring between 3 and 6 inches. Placed in pots, chokecherries are perfect for decorating patios and terraces. They grow intensely red, almost purple berries, and can be used to make jellies, marmalades, and juices. The berries can be bitter, and because they are toxic, the seeds have to be taken out during preparation. Planting chokecherries is a good way to support the local wildlife, as birds and other animals are eating the berries and searching for shelter in the shrubs.

The chokecherry is a big shrub, and it can grow to 20 feet long. It doesn't grow as high when it's planted in containers and can grow up to 12 feet tall. When planting chokecherries, make sure you put it in an area with ample space to expand. It not need much water and can grow in any kind of soil.

Instructions:

With good drainage and adequate sunlight, plant your chokecherry in a wide container during the fall. For this plant, you can use any form of soil; just make sure it isn't too acidic or too alkaline.

Initially, keep the soil moist, but less often as the shrub starts to produce quickly. Not much water is needed on this farm. Often plants obtain minimal nutrients in containers, using some compost fertilizer within the container to help your plant be healthier and grow more fruits. During winter, be patient, your chokecherry will bloom in the spring.

How to plant chokecherries:

During the spring, the black chokeberry plant produces beautiful white flowers, a large amount of dark purple, nearly black, dropping berries, and lovely dark red foliage that accentuates the beauty of fall. Its berries are sour, but they can be used to make delicious preserves, jams, and juices filled with antioxidants.

The black chokeberry is a plant which grows to a height between 3 and 12 feet. For this reason, it is stated that you pick a small plant variety for container gardening. It grows easily in any soil type and needs much water. Seeds of black chokeberry can be planted during fall. You can plant them as sprouts in spring, and you can take a cut of a sprouting root (rhizome) in the early summer.

Instructions:

One black chokeberry is planted per jar. These are plants that need their room. Make sure the container is sufficiently wide and has good drainage. Place the container for at least part of the day in direct sunlight. You can use any kind of soil for container gardening to maintain the soil moist but not damp enough to create mud or pool on the surface. A lot of water. When you want your plant to grow plenty of berries all day long and water often in direct sunlight in containers, plants often get little nutrients, use some compost fertilizer in the container to help your plant be healthier and grow more fruit.

How to plant winterberry?

The Winterberry is a shrub that fits well into container life and gives beautiful white shaded flowers ranging from cream to light green. It flowers between April and July, immediately after the plant has provided new leaves. Towards the end of the summer, the Winterberry gives around 1/4 of an inch (~0.6 cm) of spherical berries, which, when mature, turn an orange and/or scarlet red color to remain on the plant even in mid-winter. The Winterberry Berries are food for more than 40 bird species, as well as many mammals. They are, therefore, poisonous to humans and should not be eaten.

The Winterberry can measure in heights from 1.5 to 5 meters. Rather than one shrubs should be planted when planting, as both male and female shrubs are present. Planting them next to each other will help maximize pollination. They can be planted in direct sunlight or in more shady places; however, it is important that they have enough water, and that the soil is very neutral.

Instructions:

Plant the Winterberry in a wide container with good drainage and position it somewhere where at least a few hours of direct sunlight can be provided daily. The Winterberry grows well in dry, humid soil. Water it regularly, but do not allow mud to form, or collect water on the surface.

The plants have little nutrients in containers; adding some fertilizer or organic soil in the container will help your Winterberry be healthier and yield better fruits.

Seeds:

- **Sunflower**

- **Coreopsis**

- **Little bluestem**

Below are some plants which grow well in containers and produce bird-attracting seeds.

How to plant sunflowers?

The Lemon Queen Sunflower, a dwarf Sunflower variety (Helianthus annuus), is a good container planting choice. The sunflower is hardy and easily resists drought once it is formed (needs plenty of water beforehand). Not only does it survive a range of environmental conditions, but it also provides an abundance of seeds that are ideal for birds, other pollinators, and even humans! Additionally, the sunflowers have stunning, large flowers that make your home a great addition. Sunflowers grow best in maximum sunshine areas, blooming in the summer or early fall. We tend to flower well in long, hot summers, but can also grow well in areas with shorter summers. The Lemon Queen Sunflower's mature height is around 5 feet (1.5 meters), which is smaller than the typical sunflower variety.

Instructions:

Transplant them into larger pots when seedlings are several inches tall, and are at least 12-18 inches long, thereby ensuring they are spread out. In other words, no more than one plant per pot of 8 inches (diameter), or three plants per pot of 15 inches. You should hold the seedlings in wide planters 6 inches apart.

Hold soil moist and weedy. Cover the seedlings with netting or plastic berry baskets from hungry or breeding birds.

Place containers in an environment lined with direct sunlight.

How to Plant Lance-leaf Coreopsis?

The Coreopsis Lance-leaf (Coreopsis lanceolate) is beautiful and tolerant of many types of soil and environmental conditions, making it a common choice among home gardeners. It's often found in hot, infertile areas, naturally. This plant has bright yellow flowers with a diameter of 1-2 inches and dark brown seeds that are winged and curved. Often, coreopsis is called tickseeds because their seeds look superficially like ticks! This also produces plenty of seeds, which attracts pollinators. The coreopsis Lance-leaf prefers full sun but can tolerate light shade. The plant should grow fewer flowers and taller stems in areas with a heavier shade. As already mentioned, this plant is tolerant of many types of soil. This does prefer well-drained soil (preferably not clay), though.

Instructions:

Plant seeds in a bottle, directly. Seed depth can be up to 1/8 of an inch from the soil level. This plant requires reasonable quantities of water. Using well-drained soil. Using fertilizer whenever possible. When spending flowers, make sure they're pruned or "dead-headed" so the plant can start to bloom!

Put the container where it gets plenty of sunlight.

How to plant a little bluestem?

The Little Blue Stem (Schizachyrium scoparium) is a tufted herb in which the ground birds love to stay! This is often consumed by songbirds, gamebirds on the upland, and occasionally grazed by livestock! It is also widely used as a cover crop and as a prairie restore plant for erosion control. Based on the soil quality, this grass grows to a medium height ranging from 18 inches to 3 feet. For Little Blue Stem plants, the desired soil condition is well-drained, medium to dry. It tolerates flood conditions in no way very well. The growing season for this herb begins in late spring before the first frost kills.

Instructions:

Prepare a seedbed by ensuring no weeds are present and that the seedbed is solid. During spring or fall (dormant seeding), plant the seed 1/4 inch into the soil. Little Blue Stem seedlings are amazingly resilient, and in no time do you have plenty of newly developed plantings. Should not apply fertilizer to your soil the first year, unless soil tests demonstrate serious potassium and/or phosphorus deficiency. If this is the case, don't add nitrogen fertilizer– it will also allow weeds to spread.

Nectar:

Some plants also produce nectar, which attracts various species of birds, such as hummingbirds.

How to plant trumpet creeper:

Trumpet Creeper, also known as Trumpet Vines or Trumpet Flowers, is a type of American-born deciduous or partially evergreen vines. The stems with glossy, dark green leaflets can grow up to 12 m long. Blooming in July and August, the vivid yellow-orange to red flowers come in clusters of four to a dozen. Contact with this plant's leaves and flowers can, however, lead to an allergic reaction! Moreover, it is highly poisonous when swallowed. The beautiful and abundant flowers of this plant produce nectar in large quantities which attract butterflies and Alibris, making it perfect for garden decoration. The Trumpet Creeper grows with a wide pH range of 3.7 to 6.8 in wet to dry soils and types of sand, loam, or clay soil. Better blooming happens when the vine has full sun exposure, so make sure to keep it out of the shade!

Instructions:

Cuttings are usually reproduced as trumpet creeper. It readily grows roots and new suckers, making it easy to grow quickly.

You can also use seeds if you don't have a parent plant to take cuttings from.

Seeds are prepared for germination by stratifying them at 4 ° Cover 60 days in moist sand. The seeds are sown for spring out planting in early fall. The vines should be thinned during the growing season and cut back to avoid vigorous spread before winter. This plant is known as invasive weed in some areas which can wipe off native species.

How to plant red columbine:

The Red Columbine commonly referred to as the wild honeysuckle, is a sort of popular perennial herb garden. Much of its popularity among gardeners stems from its durable nature, ability to regenerate by seed, the lifespan of 3-5 years, and attractiveness to many forms of pollinators, including Alibris and bees. It's not only useful to wildlife, but even human beings have found many applications for it. In specific usage are the seeds and roots. Headaches, sore throats, stomatitis, heart attacks, skin rash/itch induced by poison ivy, kidney & urinary issues, and even fever may be handled with the seeds. The roots can be used for treating gastrointestinal diseases. The sepals are, therefore, what gives the plant its distinctive red color. The flower itself faces down and blooms from March to July, bearing fruit in the middle to late summer months.

During her first growing season, it does not bloom. When the seeds in the follicles turn black, it means they are mature and ready to collect. Red columbine prefers well-drained and loose soils with moderate acidity. This may, therefore, expand in other loose mediums and have a mixture of organic matter.

Instructions:

Seeds can be harvested from the plant between August and October at any time. Once the seeds have been collected, they have to dry for two weeks. The seeds may be stored in plastic bags for three weeks after the drying process is complete before they are ready to be sown. In greenhouse temperatures of 70 ° to 75 ° F, seeds should be sown into germination trays, which are kept evenly moist. Do not fertilize the soil, and this may damage the leaves. After 3-4 weeks of root growth, seedlings can be transplanted into plug-cells. The Red Columbines will be exposed to frost-free natural temperatures approximately two weeks before out planting (which will take place in the spring). These plants do best in partial shade but also thrive with filtered light. Offer them plenty of space to develop and expand– works well along a fence or wall. Red Columbine grows to around 1-2 feet tall.

What kind of container can you use?

Any containers capable of carrying soil would do. Many people tend to buy specially made containers for potting plants, but you certainly don't have to. Here are some budget-friendly, innovative container gardening ideas which will enable your garden to stand out. Reusing containers, you find in your home or group can reduce waste and benefit the environment.

Landscaping for birds

If your goal is to attract birds, it is important to plant berries and other plants that provide nectar and seeds. Below are some more tips on how to attract these lovely animals to your backyard.

You are already providing much-needed food in planting berry/seed/nectar plants for a lot of bird species that might be around your neighborhood, but birds need more than food to live comfortably.

It's essential to have a water source for drinking and bathing purposes. A safe way of supplying birds with water is to have a birdbath. Birdbaths are readily available in supermarkets, but there are some amazing DIY bird baths on this page that can be designed at a low cost using materials that you might already have at home! Creating your own birdbath can be an exciting family activity. You can also customize your backyard decor to match your needs! Explore more ideas here to build and care for birthplaces.

Birds like shelter, as well. Some simple and easy ways to provide shelter are by leaving a pile of deadwood (this is also simple as it offers a place for insects to grow and multiply, and birds will feast on these creatures), building a pile of brush, or just generally leaving patches of weeds in your yard. Birds will love the shelter, and soon a great variety of birds will pay you a visit if you obey all the other suggestions.

3.2 Benefits of Container Gardening

For several reasons, container gardening is an enticing alternative to ground-based gardening. Some people are attracted to gardening among plants for the very purpose of pottering around. Others, depending on whether they cultivate edibles or ornamental, look for the functional outcome of their efforts or provide for the aesthetic appeal of flowering plants. Whatever your inspiration for growing plants, container gardening takes you beyond the constraints of time and space and lets you enjoy the creative process throughout the year.

You can garden anytime:

Seasoned outdoor gardeners frequently claim they get an unrelenting itch to begin planting season. Depending on where you stay, the time will vary. It is usually at the start of winter in temperate areas and the first stirrings of life under the frozen earth. But in the tropics, when the fresh scent of earth pervades the air after the first shower reaching the sunbaked earth, it is typically the beginning of the rainy season, but those who love container gardening do not need to wait for any of those external signs. There's no need to wait for the super warm weather to begin young plants; by using a container, you can establish the ideal growing conditions.

There are no space constraints:

You may think a veggie garden doesn't have enough space. While people living in townhouses and condos can have no outdoor space to call their own if you turn to container gardening, that shouldn't be an issue at all. The availability (or non-availability) of yard space does not limit the container gardens. You don't need a yard, either. A large number of plants can be found growing on a balcony or window sill, or at a bright spot near a window. Keep in mind that several different plant varieties can be grown in the same jar. Companion planting in containers is a common concept that produces high yield and makes it possible for most people to plant gardening.

Ideal for novice gardeners:

Outdoor, in-the-ground planting is synonymous with such rough realities. One is for weeds. Their seeds are everywhere, and they germinate faster and grow stronger than the seeds we plant. That is why experienced gardeners give the preparation of veggie beds a lot of importance, time, and energy. When people take it easy and with minimal planning plant their garden. This is incredibly disappointing for newbie gardeners in particular. Natural plagues, illnesses, and vagaries are only a few other variables that can adversely affect outdoor gardens. Many inexperienced gardeners will be put off by their failure and will never attempt to garden again. While container gardening is not without its risks, comparatively, they are few. The problem with weeds is small, and diseases and plagues are readily detected and easily remedied. Being portable, containers can be transferred to safe locations when the possibility of prolonged foul weather is present.

You can bring the garden indoor:

There are hundreds of houseplants happily flourishing indoors next to a sunny window. When you have the right conditions, you will grow fruits and vegetables also. If your windows do not have the right form of exposure to allow enough light, artificial lighting can come to your rescue. There are several different lighting choices adapted to the various flowering and leafy plant needs. For example, on compact plants, the cool white light in the blue spectrum produces lush leaf growth. Ideal for the indoor growing of leafy vegetables.

In the orange-red spectrum, warm lights, on the other hand, promote flowering and compact leaf development. The tomatoes and capsicums are perfect for them. Under hi-tech lighting arrangements, you can grow vegetables that can be tweaked periodically to match different stages of growth. This allows for maximum yield from a given area. It is a blessing for those who are mobility-challenged or allergy-prone to be able to put in the garden indoors. Inside the comforts of their house, they can enjoy their favorite sport, and even fully remove soil that can harbor disturbing dust and mold.

Enjoy no-till gardening:

Even a die-hard outdoor gardener would admit its intense, back-breaking work to till the field. Additionally, tilling has been found to disrupt many of the natural organisms required for a healthy garden. It is for this reason that many people turn to a no-till garden option. Container gardening enables you to build an acceptable growing environment, teaming up with healthy components, without having to alter the soil or worry about tilling damage.

Save on water:

Planting needs more water in the field than planting containers. Also, if the root zone of plants is carefully supplied with water, much of it spread to the surrounding soil. Evaporation from a wider area easily dries the soil out and requires more frequent watering.

Plants in pots and tubes need much less water as the loss of water by evaporation is negligible because it only occurs from the top layer of soil. The downside is that potted plants only have limited access to water, unlike soil plants whose roots can expand deep into the soil and get water from lower soil layers. So they need careful soil humidity test. You may normally put your finger in the soil – about 1/4 inch down and test the amount of moisture.

Save on fertilizers:

Once it comes to feeding, plants grown from containers need less regular fertilizer applications. Just like with water, fertilizers were added to potted plants--whether chemical or organic, they last longer as they stay contained inside the containers in the small amount of soil. Potted plants do not need to share the fertilizers with competing weeds, either. For the same purpose, fertilizers should be used sparingly in containers compared to garden beds, should the roots be burned at large concentrations. Be sure to pick an organic fertilizer of good quality, or use your own compost. You'll be saving on fertilizer and having more value for money.

Pest control is easier:

Pest infestations in garden beds also need spraying of pesticides, since individual plants are not available. It is easier to manage pests in container-grown plants and does not require a chemical application.

To get rid of aphids and scale insects, you should handpick the bigger insects and use a toothbrush, or cotton bud dipped into rubbing alcohol. Another excellent way to get rid of many insect pests that inhabit the tender parts of the plant is to move the pots to the bathroom for an occasional shower. You may also take individual pots, and dunk them to drown unwanted soil species in tepid water. Holding the pot filled with water in a plate will prevent ants from reaching your pot and starting aphid farms on your valuable plants. Diatomaceous soil around the container will help to keep slugs away, as well as other soft-body pests.

You can adjust the height of your garden:

If you have any physical disabilities that hinder your mobility, or simply don't want to bend to the earth, the solution is container gardening.

You can have wide tubs at a comfortable height, or place pots for easy access on a ledge or shelf. It makes it easy on your back for chores like watering, feeding, and deadheading. By holding the plants at different heights, you can optimize the use of space.

Harvesting is a breeze:

If you grow vegetables and fruit in containers, it makes it much easier to harvest. Grow in pots, not just strawberries, and blueberries but all kinds of root tubers such as carrots, radishes, potatoes, and sweet potatoes. Rather than digging them up and doing any unintended harm to the precious produce, just overturn the pots on a plastic sheet when they are ready for harvest. Shake the soil out, and you'll be in perfect shape for every single tuber.

You can adjust the growing condition:

The right climate for your container plants is easy to have without performing large-scale soil modifications. You can provide your rhododendrons and blueberries with a slightly acidic source, without upsetting the soil's pH in other containers or garden beds. If the light intensity variations are seasonal, you can rearrange plants accordingly. Plants with similar watering needs can live in harmony for as long as they are housed in various containers.

You are free to choose the growing medium:

Container gardens let you play with various through media and techniques. For starters, you can do away with the soil completely and grow your veggies in an inert medium such as expanded clay pellets intermittently bathed in the nutrient solution. This hydroponic culture improves yield and prevents diseases usually occurring in garden soil caused by pathogenic bacteria and fungi. Many alternatives include the coco peat and sphagnum moss.

You can do an instant makeover:

Container gardening lets you change your garden's look and theme by simply changing the containers. Large stone containers in various architectural forms, for example, can give the garden a traditional look and feel, while metal containers with copper patina or rust – whether true or fake – will take you to a different time and era. Acrylic containers in jewel colors can make your garden a fun place, but sophistication can be spelled by a monochromatic scheme with the same content.

Add seasonal color to the interior and the garden:

Seasonal plants such as holly, Christmas actus, and poinsettias just look nice for a short period of time. Make the most of their brief show by showing them throughout the season in prominent roles, and then push them away to obscure spots.

Brighten up the dark corners:

Plants with variegated gold and silver leaves are perfect for brightening up dark and dingy corners. Under these low-light conditions, they cannot live for long, but after a week, you can switch them to a sunny location and get other plants to fill the dark spot. It helps in getting many rotational plants.

How to get started with container gardening?

Container gardening is easy when you just keep in mind the basic plant requirements. All green plants need sunlight, water, air, and a growing medium; the first three to promote their food-producing photosynthesis, and the medium to anchor the plant and provide nutrients.

Sunlight:

Various plants need different quantities of this. Ideally, those who enjoy full sun will live outside all day, whether they are in containers or not. Many flowering plants and most fruit-producing veggies are lovers of the light and need 5-6 hours of direct sun to do their best. In partial shade, most leafy plants, including leafy vegetables and those with ornamental foliage, do well. They do well indoors if they receive direct light for at least 3-4 hours. They, too, will thrive under artificial light.

Many plants can live in the full shade, while bright indirect light will be appreciated. They make plants suitable for indoor use.

Water:

It is important for all their metabolic functions, so daily watering is a must for potted plants as there is no other way to get it from their roots. That said, most plants dislike waterlogging because it drowns their roots literally, and causes them to rot. This is able to destroy a potted plant in no time. Some plants such as cactus and succulents store water in their tissues, enabling them to go for longer periods without water. They're great for beginner gardeners who can fail to water them regularly.

As a general rule, overwatering is better when it comes to container plants under irrigation, but they will be the happiest if you test the soil for the amount of humidity before irrigation.

Air:

Plants need a fair amount of air circulation, not only around their crown but also in the root region. Growing plant cell needs to breathe, so they can be suffocated by compressed soil and air stagnation.

Growing medium:

A growing medium can be sand or perlite-modified garden soil, to allow good drainage.

The application of compost or leaf mold ensures adequate water preservation and availability of nutrients. Container plants grown in soilless media require a steady supply of nutrients, and frequent flushing out of fertilizer toxic by-products. If it contains soil or not, the growing medium should be porous enough to allow good drainage and air movement.

Some plants, like philodendrons, may be grown in containers filled with water. While they live fully submerged in water with their roots, they certainly do better in a hydroponic system that allows air circulation around the roots.

Choosing containers:

You can grow plants in any container that can accommodate enough growing space, but holes in drainage are a prerequisite for this. They allow the circulation of air, as well as excess water drainage. Containers made of glass, plastic, metal, concrete, ceramic, and stone are common choices.

The walls are porous with clay containers. Although they house the root zone in cool comfort, the soil dries out faster, and more frequent watering is needed of the plants. Plastic, metal, and ceramic pots do not allow their sides to evaporate, so care should be taken to ensure proper drainage and careful watering.

Choosing plants:

When you start with container gardening, it is best to choose strong, easy-care plants. Chinese evergreen, dieffenbachias, snake plant, philodendrons, zee plant, dracaena, cast iron plant, and ponytail palm are perfect options when you decide to grow them indoors. African violets, spathiphyllum, orchids, and anthurium do well indoors, among flowering plants, but can need some extra care.

You have the endless range to choose from for the outdoor container gardening. Quick all kinds of vegetables and dwarf fruit trees can be grown in containers. Since the containers can be transferred into growing tents or indoors during winter, the growing season can be extended.

As you become more experienced with container gardening, you can move on to groupings of plants that may be of great visual appeal. To grow together in the same pot, you can choose a number of plants with similar cultural requirements. A mixture of purple coleus, chartreuse sweet potato, and pink and white petunias, for example, make this a striking arrangement.

Feeding:

Container plants should be sparingly fertilized, and only when they are in the active growth stage. Accumulation of fertilizer residue can be a concern, particularly when using non-soil media because there are no soil organisms to break down the toxic by-products. It may be appropriate to periodically flush the media under running water or to immerse the pots in large containers.

Re-potting:

Re-potting is one additional task in container gardening. If you grow houseplants or annual plants and depending on the root growth of individual plants, this must be performed once every two to three years. If water drains out too quickly, or a lot of roots come out through the drainage holes, repotting might be appropriate for the plant.

Gently ease it out of the pot after watering the plant by pulling and twisting at its base. Remove the dead roots and soil as soon as you can, without harming the healthy roots. Repot in a two size container larger than the current one. Securely repair the plant in the medium and leave it in the shade until the plant regains vigor.

Soil addition/replacement:

If you grow annual veggies seasonally in containers, the soil will need to be added or replaced before it grows the next year again. Be sure to fully remove the soil if you find any signs of disease in the plants during the previous year because it may linger in the soil.

3.3 Steps to Grow Plants in Containers

Here are some basics steps for growing plants in the container:

Choose a container:

Seed starting containers should be clean, should be at least 2-3 inches deep and should have holes in drainage. They may be plastic bottles, packs of cells, peat pots, plastic sheets, cups of yogurt, and eggshells. The options are infinite, as long as they are safe (soak up 10 minutes in 9-part water to one-part household bleach).You can buy seed starting kits as well, but don't spend a lot of money until you are sure to start seeds each year. When you start seeds in very small containers or plastic sheets, then you may need to transplant seedlings into slightly larger pots until they have their first set of true leaves. Bear in mind it takes up room for flats and pots so make sure you've got enough space for all the seedlings you launch.

Start with good quality soil:

Crop seeds are available in sterile, crop starting mix or potting soil in nurseries and garden centers. Do not use garden soil, and it is too thick, it contains seeds of weeds and, probably, species of disease.

Plant at the proper depth:

On the seed packet, you can find the correct seeding depth. The rule of thumb is to cover soil seeds equal to three times their thickness. Some seeds need light to germinate, like some lettuces and snapdragons, and should rest on the surface of the soil but still be in good contact with moist soil. Smooth tamping will help after sowing. Using a spray bottle after you have planted your seeds to cool the soil again.

Water wisely:

Using water at room temperature indefinitely. Let chlorinated water sit overnight to dissipate chlorine, or use purified water. Consider using water which is salty. Keeping soil consistently moist is necessary, but avoid overwatering, which promotes diseases which can kill seedlings. Try not to splash your leaves with mud. An easy way to prevent this – as well as overwatering – is dipping your container base into the water and allowing the soil to retain moisture from the bottom up to the surface. Some seed-starting kits include a wicking mat that leads water from a reservoir to dry soil. It may be the most goof-proof way of watering seedlings, but you have to be careful not to leave the soil too warm. Whatever you do, don't forget to water and let the seeds or seedlings dry. This is sentenced to death.

Maintain consistent moisture:

Cover your jar before germination, to help trap moisture inside. Usually, seed starting kits come with plastic cover. A plastic bag may also be used, but it should be protected so that it does not lie flat on the surface. Remove covers as soon as seeds germinate. Once seedlings grow, reduce soil watering partially dry, but don't let them wilt.

Keep soil warm:

Seeds need to germinate in moist soil. Most seeds are to germinate at around 78 ° F. Waterproof heating pads, specially built for seed germination, maintain the soil at a steady temperature. Many nurseries and garden centers sell purchasing them. Alternatively, before the seeds sprout, you can put seed trays on top of a fridge or other warm appliance. Air temperature will be just under 70 ° F after germination. So long as the soil temperature stays 65-70 ° F, seedlings can tolerate air temperatures as low as 50 ° F.

Fertilize:

Start feeding your seedlings after growing their second set of true leaves, applying a weekly liquid fertilizer of half power. Gently apply it, so that seedlings are not dislodged from the soil. Apply a full-strength liquid fertilizer every other week after four weeks before transplantation.

Give seeding enough light:

Not enough light leads to leggy, tall seedlings that will fail outdoors until transplanted. You can grow stocky seedlings in a luminous south-facing window in mild winter areas. Farther north, even a window facing south, may not give enough light, especially in the middle of winter. Ideally, for healthiest growth, seedlings require 14-16 hours of direct light per day. If seedlings start bending towards the window, that is a sure sign that they do not get enough light. Turning the pots just won't be enough-you may need to provide artificial lighting. Lighting kits can be issued by nurseries and mail-order seed catalogs. Follow the directions with care.

Circulate the air:

Circulating air helps prevent disease, and allows strong stems to grow. Run a gentle fan near to seedlings to build a movement in the air. Hold the fan away from the seedlings, in order to prevent overt blasting.

Harden of seed before planting:

Hardening-off seedling is a process that should be carried out slowly over a week or two. Done properly, you will produce sturdy seedlings ready to stand up to anything that nature throws at them. Rush the process, and the young transplants may be burned, hurt, or even killed.

Harden-off tender seedlings over a 6-14 day period. The goal is to slowly expose transplants to environmental conditions, increasing the number of time seedlings spends each day outdoors. In the process of hardening-off, slowly minimize watering. You don't want to wilt seedlings, just extend the time slowly between watering. Stop feeding for 3-4 days before starting to harden. Do not feed before you transplant yourself into the garden again.

Start by growing seedlings in a sheltered spot outdoors-shielded from wind and direct sunlight. This is ideal under a branch, or overhang. The first two nights put seedlings back inside. Place them out to another 30-60 minutes of brighter sunshine per day. Work your way up to give more and more direct morning sun to plants, until they can withstand midday sun without getting wilted. Fully hardened-off seedlings should be able to endure the same amount of sunlight they get in the garden as they are being cultivated.

Check weather forecasts and keep track of expected lows at night. When temperatures drop below 35 degrees F., carry indoor seedlings or cover them with a spun-polystyrene row cover or other protective material. Read more about shielding plants from frost. When you put the row cover on hoops or stakes, you'll get the best protection against frost so that it doesn't directly contact foliage. If it looks like a frost or freezing temperature will occur, prepare to leave seedlings overnight outside by the third or fourth night. Place them under a protective overhang, or under a table. Seedlings shouldn't need protection by the sixth night, or so.

Your transplants will look more stocky and tougher after about a week of hardening-off. Plant on a rainy or drizzly day when the winds are calm, or at night. Water in with diluted to half-strength liquid fertilizer solution. Provide protection against pests, such as snails, slugs, and cutworms, where appropriate. Continue shielding seedlings from excessive downpours, frost, hail, or strong winds. A plastic gallon jug with removed bottom allows for a nice seedlings cover.

3.4 Best Plants to Grow Using Container Gardening

The pots and containers give great flexibility to the gardener and are a wonderful way of experimenting with planting and design. From short-term bedding displays to permanent features of small trees and topiary, planting in pots adds a different element to the garden, softening corners, brightening dark spots and producing immediate, but easily changeable, results. Stick to a single or two different materials when choosing your bowl. Take your cue from the theme of the house, and garden-red brick houses are enhanced with terracotta containers, while the best setting for galvanized metal pots is a modern property. Bigger pots have more effect, and plants growing in them do not dry out as easily, but an eclectic collection of small containers creates an ever-changing, quirky scene. Repetition may be an efficient-for maximum impact; garden designers often use three or more similar containers planted with the same plants.

Coreopsis tinctoria:

Coreopsis looks amazing when paired with other annuals and perennials in a bread bowl. Seek to mix with phormiums, euphorbia, nasturtia, and violet basil.

Cosmos:

The cottage-garden look of daisy-like cosmos flowers gives pots and containers a sense of informality and movement. They fit well with silver foliage plants, but they are successful on their own as well.

Busy Lizzies:

Busy Lizzies are suitable for shady gardens. Regularly deadhead them to ensure they bloom into autumn.

Clematis:

Some clematis is fit for container production. Evergreen clematis is often sold tied to a stake upright, but its trailing habit makes it suitable for a container's bottom. Partner for a dash of color with spring bulbs.

Ivy:

Ivy is one of the most desirable and versatile potting plants. "Ivalace" Hedera helix has dark-green, glossy leaves with curled edges.

Euonymus 'Emerald 'n' Gold':

The variegated foliage is a real pick-me-up throughout the year. Try it with Jenny, Lysimachia nummularia, tulips, golden narcissi, or primroses for a stunning show.

Pittosporum tenuifolium:

This elegant evergreen shrub has rich mahogany leaves that emerge before darkening in a pale creamy color. Slightly tender, a sheltered spot will be needed over winter.

Skimmia japonica:

This male form is blessed with glossy evergreen leaves and an abundance of tiny red buds through the winter that opens in the spring to pinky-white flowers.

Hosta:

Hostas make beautiful architectural plants and perform well either alone or with other plants in containers. Seek to mix with heart bleeding, or other plants with foliage, including heucheras.

Fountain grass

Fountain herb is a real show-stopper. Build up for dramatic effect in large pots with alliums, or make a statement on its own. 'Rubrum' has graceful stems and red-tinged, squirrel-tail flowers which fade in autumn to beige. Cover from winter frosts.

Viola:

While typically grown as an annual, violas are frost-tolerant, and productive in many winters. Its compact size and free-flowering habit are a perfect match for gardening containers. A number of bright, festive colors are available, including white, lilac, purple, yellow, and orange. Sunset Boulevard is the name of this combination of lemon and peach colors. Violas are available at a local garden store you want.

Baby cakes blackberry:

Who wants blackberry to be used in container gardening? They're gangly and spiny! Ok, Baby Cakes is a thorn less, dwarf blackberry. The compact habit (maturing at 3-4 ft. in height and width) is ideal for pots in the patio. Throughout summer, Baby Cakes offers tall, tasty blackberries and, in some climates, will produce an encore crop later in the season. In Zones 4-8, it is hardy. Mulch the pot or bury it in cold climates for winter safety.

Purple Majesty Ornamental Millet

When it comes to container gardening, here's a plant with true stage presence. Using it alone or to back up smaller companions — just be sure to put it in a big bowl. Ornamental Purple Majesty millet grows 4-5 ft. tall and 8-12 in. Widespread. Its elegant foliage takes on a deeper purple hue in full sun to make it one of the best ideas for floral landscapes. Heat-tolerant and low-maintenance, Purple Majesty, provides interest from spring to early fall. It stays on long enough in some climates to bring value in winter too.

3.5 Maintenance of Container Garden

Maintenance of container garden is important to keep your container garden beautiful and safe. These tasks are usually simpler but essential because we have better access to the plants in our containers.

Grooming:

One area of maintenance in a container garden is grooming. Grooming is what keeps most annuals looking their best, and plenty of perennials. It encourages repeated bloom times and increases the number of blooms that a flower will create. Grooming can be as easy as the blooms expended on deadheading, or it can require cutting back or pruning plants.

If you remove dried, withered blooms, many annuals, and perennials can continue to bloom for longer. This is known as 'deadheading.' What you need to do is pick your fingers off dead blooms. Using pruning shears on rough trees, or woody roots. Cut or pinch the flower directly below it, just above the leaf. Allow until most of the blossoms are ready to harvest and then use garden shears to shear them all off if a plant has tiny blossoms that are hard to destroy individually. By doing so, several plants will flower again in a few weeks. Shearing toiling would also increase the overall appearance of the plant.

Cutting back:

Maintenance of container gardens may require cutting back. Some flowers can get too tall in their container and look awkward. Cut them back to promote more compact growth and multiple blossoms. You will also want to switch your containers around from time to time so that there is enough light on each side of the container. When they have to reach too far for the sunshine, plants start looking straggly. Cutting back also increases edible plant development, including herbs. When the plant has been set up, pinch the main stem off. From this point, it'll fill out with new branches. If you have flowers, pinch them off too. You may cut large amounts of herbs for use, drying, or freezing when the plants are big and whole.

Old-growth, stems, and leaves are usually trimmed. This many times prompt new growth. When you think a plant is blooming, then look for new growth at the plant's base. If you're waiting a little bit, you may be shocked to find it'll bloom again.

Pruning:

Pruning is a chore required for proper maintenance of the container garden. When caring for container gardens, which include larger plants, pruning is required. It encourages plant protection and improves plant shape. Cutaway growth is dead, impaired, or diseased. Excess branches, too, should be withdrawn. This will allow more air to circulate between the plants and will allow sunlight to touch all of them.

For a more natural look, cut back branches at different lengths while forming a shrub or tree by pruning. Render cuts at or only above branch unions.

Thinning a shrub or tree helps encourage blooming but does not alter the form. Which requires the elimination of whole branches? Should not leave a stub of a branch sticking out from the main stem or root when cutting off a branch. It causes sickness.

After they bloom, most shrubs can be pruned, but check out the guidelines for different plants. Some need to be pruned late winter or early spring (wisteria and hydrangea), while others need to be pruned after spring blooming, like lilac. Eventually, cutting the primary and secondary stems can produce further growth and also result in second and third sets of buds and blooms.

Don't prune into the heat of the day. When planning maintenance of your container garden, remember that early spring is generally the best time to prune deciduous plants. You will see the structure of the branch, and the wounds should have time to heal before winter. Prune fast-growing plants in season at any time, but not during the hottest part of the day.

Re-planting:

Maintenance of container gardens may require replanting occasionally. Maintaining containers to look at their best can mean replanting. This is yet another container gardening elegance. When a flower has finished its time of bloom, or a plant wilts or dies for some reason, we may simply take it out and plant something else. For a seasonal change, this can be done too. Replace African daisies, spurge, or flax with geraniums or petunias for fall. Add asters or even chrysanthemums. Keep the container complete during all seasons. . Repotting 2 to 2 1/2 times the size of the outgrown one in a new pot will save you from having to do so too much. Trim the larger roots of the plant back by as much as one-third when replanting, and make sure the roots don't wrap around the base. If so, remove the ball and repot several inches wider in a planter. Using the fresh mix of potting and apply the fertilizer.

Overwintering perennials:

Know how to overwinter them, to protect your investment and enjoy your annual flowers, trees, and shrubs. If you live in that environment where the temperatures fall below zero, hibernation is a must. There are many winter approaches to use-read about them and what would be the best work for you

Container care is part of container garden maintenance:

Forget about proper container treatment when preparing container garden maintenance. If you have pots or containers that will not be used during the winter at the end of the blooming season, wash them out and clean them so they'll be ready for the next spring. Get rid of the old potting mix should it contain rodents or larvae. Wash the bowl clean with water and soap. Dip the pot in a poor solution of bleach and water if the pot contained diseased or pest-infested plants at any time and allow it to air dry. Wintertime is also a good time to spruce up your containers and pots too. You may have white containers you want to paint, or you might have chipped pots that need to be cleaned up. Only note painting is not waterproofing. Wait until the paint has dried thoroughly, and then apply a few coats of polyurethane to weatherproof, if you paint your containers.

Pest and disease:

Maintenance of container garden means managing the pests and diseases in your boxes or pots. Like other gardeners, container gardeners have to deal with insect invasion in container gardens, and disease in container gardens as well. Since we water regularly and our plants are nearby, the positive thing is that we will regularly detect issues with insects or disease early and have the ability to fix issues before they become major infestations.

Chapter 4: Vertical Gardening in Urban Areas

A vertical garden which grows upwards (vertically) using a trellis or other support structure, instead of (horizontally) on the ground. Any plant grown on a trellis or on a fence is part of a vertical garden in the technical sense. This procedure can be used to create living screens between different areas, giving your yard or home privacy. Vertical gardens can also be used more recently to grow flowers, and even vegetables

Some use vertical gardening as a way to ensure they use their garden space to its full potential. For instance, a simple structure built by bamboo poles will allow bean plants to climb vertically, creating a more growing space than a traditional horizontal garden would have been possible with. Squash, cucumbers, and even tomatoes can be grown vertically, too.

When it comes to vertical gardening, climbing plants and vineyards are far from the only options. With a small planning and the right materials, you can build vertical gardens that allow you to grow virtually anything. One will find a variety of DIY kits that use small cups or other containers placed in rows in front of vertical support. They are packed with soil and seeds and then fed.

You're not, of course, limited to use a single row of containers set in a vertical sheet. You can also use almost any program that lets you grow up rather than outwards. This includes scaffolding, machine shelving, and more. Only build flat surfaces along the vertical axis at different intervals, and add plant trays or containers.

It is considerably easier to grow crops from a vertical garden than with a traditional on-the-earth garden. Since you can harvest while standing mostly upright or fully upright (depending on the vertical amount harvested), as opposed to sitting or squatting on the ground, vertical gardening is easier on the back and legs, and it is also useful for many people with arthritis or other disabilities.

4.1 Purpose of Vertical Gardening

A vertical garden is a technique employed by using hydroponics to grow plants on a vertically suspended stand. These specific structures can be either stand-alone or connected to a wall. Since ancient civilizations, vertical gardens have been used; many modern vertical gardens will last for decades and give the modern-day company a pop of nature. Vertical gardens in the office space provide a great alternative to potted plants. While potted plants have the benefit of being put anywhere, they can take up space and require much maintenance. With vertical gardens, however, there is only one wide panel to maintain, and it will give any professional environment a lush pop of color.

If you think you have a vertical garden in your office, maybe for a lobby, public walkway, or meeting room, then you need to contact your local Ambitus office today. Speak to one of our award-winning designers and explore what we can do to build a green wall that stands out for your workplace.

Vertical gardens are known by several different names: living green walls, living walls, and moss walls, to name only a few. Such vertical structures of plant life can be as small as a picture frame or as large as a 60 ft. long masterpiece, whatever you want to name them. Vertical gardens can be built-in hotel lobbies, the headquarters of major companies, or even a small residential backyard. Not only do vertical gardens look spectacular, but they also make you feel more relaxed in your surroundings by tapping into the intrinsic soothing forces of nature.

Not only do vertical gardens provide a wonderful centerpiece for your interior room, but they also improve the natural airflow into your setting. These gardens can be made up of several different plant types; the most common plant species used in green walls are ferns, focus repens, pilea, and calathea; Even vertical gardens are very spatially effective and can fill any empty space on a wall.

Many workers are exposed to a lot of dangerous air pollutants like formaldehyde and carbon monoxide in workplace environments. Vertical gardens act as a natural clean air network and foster a better respiratory atmosphere and a safer overall climate. Plants also help to minimize noise pollution that provides a safer working atmosphere for occupants to construct.

Many of the country's and world's buildings have vertical gardens built on their exterior. Many outdoor gardens contain moss, trees, and other plants that are commonly used in outdoor vertical gardens. Exterior vertical gardens have the advantage of providing direct natural sunlight, which will help them grow.

Outside vertical gardens also give buildings great protection and insulation from changes in temperature, UV radiation, and heavy rain. Outdoor vertical gardens use a process called evapotranspiration during the summer, which helps to cool the air around it. As the environment is vastly different across North America, plants for outer walls Ambius installations are picked by the state environment, making it easier to maintain the greenery.

Most Ambius vertical indoor gardens have an easily maintainable panel or tray framework. Panel systems such as the Sage Vertical Garden system can help you keep your lovely vertical gardens on the lobby or office walls. A tray-based device like the Vice Wall will be useful for certain ventures. To help the plants remain hydrated, this form of device uses polypropylene trays and a water-resistant back panel. When you don't have the experience to care for the structures, without proper maintenance, a vertical garden will easily fail. It includes watering the plants, as well as pruning and removing dead leaves. Most vertical garden systems have complex irrigation systems that ensure proper watering of the plants, although some degree of maintenance is also needed. It is also lucky that Ambius performs monthly updates on several different systems.

Succulents are popular with plant enthusiasts as they are spatially effective, colorful, and easy to maintain. Succulents are also fire-resistant, and if you are in an area with a lot of water shortages, they are perfect alternatives. Succulents are excellent plant options for vertical gardens, for that reason. Succulents are well known for having a hard and fleshy exterior, which makes them great indoor candidates. The range of succulents makes molding and shaping any vertical garden very convenient for your needs. Watering and other maintenance are required less for succulents while providing any room with a wonderful aesthetic.

When should you consider a vertical garden?

The response to that question depends on what you want your company to do. You can do this if you only want to have a few plants around on top of the filing cabinets and on the concrete. But, if you have a large lobby room or a large space that you like you could really use to brighten things up with some greenery, a green wall could be ideal for you. Ambius designers can help you identify areas of your business that could benefit from a vertical garden. We can also help you check out which green wall solution works best for you.

Why are vertical gardens perfect for offices?

Offices are commonly considered to have in a fairly small room a number of people packed. If you remember the studies that show office workers who have easy access to view greenery or nature to boost creativity and overall mental health, and you start to understand why it is necessary to have a green wall. Plus, when you're working with an Ambius designer, they're working with you to ensure the vertical garden really stands out and fits with the decor of your company. Would you like to have a vertical garden in living plants with your logo rising in the center of that? It is possible with a green wall on Ambius.

Green walls help to clean the air, to enhance the climate with the plants generating fresh oxygen. The green vertical garden wall is a conversation starter and has a "wow" element in holding your company top of mind. Vertical gardens are also ideal for offices because it suits many plants in a very small area and makes the office look better.

Greenery in an office or in any interior setting has a lot of advantages, but the one major advantage is the way you feel after installation. Many studies have shown, time after time, that plants have a profound impact on mood improvement in work environments. If you wish to have a vertical garden built for your room, please contact Ambius for a consultation today.

4.2 Benefits of Vertical Gardening

Due to the restricted space in your high-rise apartment, does your passion for gardening fad away? Allow no worries. The solution to your dilemma could be vertical gardening. Very many businesses invest in the vertical garden for their architectural, physiological, economic, and ecological benefits. This style of gardening is also ideal for gated community multi-story apartments in Chennai, which is very popular in metro towns.

Save Your Space:

As already mentioned today, most of us don't have outdoor areas to go to a garden, people are living in an apartment and could only offer a small area of their balcony for gardening, so they should be very creative to make the most of it.

A vertical garden uses very little planting space, and you can have piles of plants arranged vertically one by one, or hanging one above the other. You have to be careful that the plants get the proper sunlight.

Keeping your plant diversity:

You can also play with can plant diversity, such as decorative and vegetable plants, row by row to give the garden aesthetics. The vertical garden is also easier to grow and maintain compared to the others.

Healthy gardening method:

Plants are grown off the ground that minimizes the risk of pests. It also protects the plant from harm from pets digging up the outdoor gardens. Where to start a vegetable garden in your apartment, too? Get your green day started as quickly.

A Protective Shield:

Everybody knows that plants absorb toxins and toxic chemicals, so it serves as a defensive shield when we grow plants vertically in a compact way and gives you a pure atmosphere to breathe in.

Insulate your building:

Not only does your green living wall absorb the pollutant from the soil, but also heat, noise, harsh weather, and UV rays. It controls temperature through transpiration and gives you refrigeration.

Simple to maintain:

Simple to reach vertically arranged plants – also makes fertilization, irrigation, pruning, and harvesting convenient.

Gives you privacy:

You can expand your green wall outside windows or a little away from the door but keep it concealed from the outsider, and it also gives you shades out of the sunshine and protection from the passer-by.

Living room divider:

On wheeled containers, you can grow a vertical garden and keep it indoors as a beautiful & attractive room divider. This wheeled container will help you push as needed the garden into the sunlight.

Reuse your waste

Using recycled materials such as plastic bottles, old shoe organizers, a broken ladder, basketball, bike, can, etc., can easily render this form of green architecture. Grow More Plant types: You can literally increase the plant varieties by adding to the top of it the climbers like cucumbers, tomatoes, and melons on the downside of the garden and the small plant varieties.

Bad soil quality? Don't worry:

If you have low soil quality in your garden, there's no need to go for cost-effective treatment to expand your garden; you can always opt for vertical gardening using good quality soil.

A healthy vegetation:

The design gets more exposure to sunshine and air circulation in vertical gardening as it grows upwards, which results in a healthy garden.

Plants reduce stress:

The physiological role of plants in reducing physical signs of stress and encouraging exposure to vegetation has been found in several studies to have a greater beneficial impact on human health.

Aesthetic Visual Attraction:

Vertical gardens on the wall are ivy. With vertical gardening, you can amazingly turn the wall or any empty space into something esthetically pleasing and creatively stimulating. It can be used to render green wall, or gate decoration, or hangings basket or window boxers, whatever it maybe gives a natural beauty that is calming and has a great impact on the visualization. This will help you turn your apartment into luxury villas that you always like.

Living green walls purify the air

Plants greatly enhance the air quality in a living environment. The wall traps airborne particulate matter, turning CO_2 into oxygen. Just one m2 of the living wall absorbs 2.3 kg of CO_2 from the air per annum and produces 1.7 kg of oxygen. Many studies have shown that plants and microbes present in soil media accumulate harmful VOCs and turn them into a compound used for food by plants.

Green walls increase the sense of well-being well

A cleaner air leads to fewer health problems, such as headaches and respiratory irritations, as well as greater attention and concentration. Complaints like tired eyes, headaches, sore throats, and weariness decrease. A significant reduction in absence due to sickness exists in workplaces where there is plenty of greenery.

A living wall reduces the ambient temperature

Radiation is absorbed by plants. More precisely, it consumes 50 percent and reflects 30 percent. This helps create a warmer, more-friendly summer atmosphere. This also means that 33 percent less air conditioning is required for the indoor environment, which in effect means energy savings.

Vertical gardens minimize ambient noise

A living wall serves as a built-up sound barrier. It absorbs 41 percent more sound than a typical surface, resulting in a decrease of up to 8 dB. Thus, the world is much quieter with noise levels comparable to those present in nature, both inside and outside the house.

Green walls alleviate tension

Living in a green environment has a beneficial impact on people's health. Green workplaces promote relaxation by reducing tension. The blood pressure, heart rate, muscle tension and brain function could all be enhanced by watching plants for as little as 4 minutes.

Vertical gardens improve efficiency

Directly linked to the sense of well-being, positive moods enhance learning and make decisions on complex tasks more effective. Green exposure also results in greater use of rational thinking and more creative methods. A green workplace will lead to a productivity increase of 15 per cent. Plants affect employee satisfaction positively.

Green spaces demonstrate that workers

The company is not the only one to enjoy the advantages of a longer residence period. Green walls often create beneficial externalities for people who don't live in those areas or spend much time. Just like store or restaurant customers. In reality, they can only enjoy the benefits by making the decision to spend time in that particular room. In this latter case, the longer, on average, residence time customers spend in the shop and the greater satisfaction with the customer experience records one beneficial result.

Plant walls add value to the brand

The natural and organic appearance, combined with a reduction in the cost of energy, means an increase in the value of the land. Studies have shown that the construction of a business can be seen as a sign of its environmental and social success, and can be an attraction for job applicants.

More social contact, less vandalism

It brings people together to work or live in a green environment. Small scale greenery, in particular, has been shown to have a positive impact on social cohesion in neighborhoods. Areas with more greenery are suffering less from aggression, violence, and vandalism in that respect.

Living walls offer a powerful "healing environment."

Greenery encourages patients to recover more quickly, resulting in a shorter stay in the hospital. In a green area, a person's tolerance for pain is higher. Also known as 'healing setting' is this powerful influence. Through actually looking at areas dominated through green plants, flowers, and/or water significantly fosters regeneration and decreases patient tension.

4.3 Steps to Grow Plants in a Vertical Garden

There are two choices if you've wanted to create a vertical garden: either you can hire a specialist company to come and install one for you, or you can set up one for yourself. The advantage of paying professionals to do the job is clear; they have the experience of building these gardens, and they will know exactly what is required. You may be assured that the garden will remain in place and will function as it should. These installations are not cheap, though, and you will want to try to set it all up yourself. This is a much cheaper alternative, and the good news is it isn't that hard. There will be a certain initial outlay for the content, but you stand to save quite a bit of money by doing so yourself. Essentially, a vertical DIY garden would need a frame to keep it up, a plastic sheeting backing layer, and a cloth layer for the plants to live in.

Start Choose the wall for the plants

Choose the wall before you even start constructing your vertical garden; you need to take a moment to consider the outcome you're looking for. Specifically, you need to determine whether you're selecting plants for a wall or a plant wall.

In other words, is your primary concern covering a specific wall, or is it your main objective to grow other plant types? If your aim is to cover an unattractive wall with greenery, you will need to choose the plants that you place there according to the location; on the other hand, if your aim is to grow specific plant species, you should find a location that will best allow them to thrive.

If you want to grow sun-loving plants, for example, pick a wall that will give them as much sunlight as possible. If your selected plants prefer the shade, you can choose somewhere that is shielded from the sun's direct rays.

Often consider factors such as wind, rain, and temperature – as well as whether growing a particular species where you live is really practical. Understanding how plants grow in nature will help you choose your vertical garden's best plants.

Set up the frame

You need to create the frame until you know what kind of plants you should grow or where you want to put your vertical garden. When it comes to content, you have a few options here. A metal frame will be strong and sturdy but also heavy, not to mention costly. If you are considering a metal frame, consider carefully whether the wall to which it will be attached can support its weight – particularly once the rest of the garden is constructed and hanging on it too. Another choice is the wood-but note that as you water your plants every day, the wood will be continuously wet. This means it will quickly rot if it isn't treated properly. PVC plastic piping probably is the best choice. It would be cheaper and lighter than a metal frame, and there will be no problem bearing the weight on your wall. Also, it will last longer than a wooden frame because it does not respond to constant watering. To start with, just build the frame, but don't add it to the wall yet – you still have to repair it first on the other components.

Proceed. Attach the back layer

The next step is to connect the back layer to the frame, which is a layer that serves a dual purpose. This will mainly form an inner layer at the top, behind the cloth. You will then build pouches for your plants between the outer fabric layer and the inner layer when the frame is completely built. Secondly, it will also help protect the wall against the water that continuously flows down your vertical garden when you water it every day. Any type of hard plastic sheeting would make the best layer to back up. You want to get something that can stand up to a minimum of rough care without tearing.

Attach the layer of fabric

Next, you need to add the layer of fabric to the front; this will be the visible part, and this is where you will put your plants. For this, you can use any material; the only prerequisite is that it can hold water without becoming rotted. If you want a professional-looking wall, special materials can be bought for this, but if it is later covered in plants, there is not too much point in spending lots of money on it. If you don't want to purchase expensive items, you can either use cheap fabric or felt. Fix the fabric onto the mainframe to ensure that the fabric is extended without creases. The best approach is to have the fabric stapled or screwed in place.

Install irrigation and fertilizer systems

You can soon learn that supplying water for a vertical garden is something of a challenge, unlike a normal garden. You might probably water it with a garden hose, but installing an irrigation system at the top is a much more effective and labor-saving way to do it. You can configure the garden to water itself, using an irrigation system. It'll probably be best to make multiple short bursts per day, which would be quite difficult if you had to do it manually. Irrigation devices with exact timers can be ordered from the suppliers of irrigation. If you'd like to make a DIY version, you might create it by drilling holes in plastic piping so that water drips down to water the plants – but you'll still have to manually push the water in. You'll also need to put in a system to provide fertilizer to your vertical garden when you've settled on an irrigation system. The Fertilizer injection systems are also available for purchase, which are well suited to the purpose.

Connect the frame to the wall

You need to install it on the wall until your frame is assembled and ready to fit. You will decide to hang it or build it permanently, so it's easy to detach. A portable garden has the advantage that it helps you to grow warm-weather species; in the colder months, you can simply take down the garden and place it cooler elsewhere, like in a garage. If you do a permanent installation, however, it will likely be better able to withstand more extreme weather without falling off the wall. It also depends on how big your vertical garden is-eliminating larger gardens for winter storage cannot be practical.

Hooks will suffice for a temporary version; you may choose to hang the garden on brackets for more permanent installation.

Attach plants and get imaginative now

Your vertical garden is over. All that remains is to have the plants added. Make small slits in the outer sheet of fabric with a sharp knife to create pouches for your plants. Extract as much soil as you can from the roots of your plants, and put it in the slits. Once each plant is in place, make the pouches around the roots of each plant by stapling the outer layer of fabric to the plastic backing. Try to keep the staples as clean as possible, but don't think too much about it, as the leaves of the growing plants will cover them soon.

This is the simple method of creating a vertical garden-now your canvas is your vertical garden. Experiment with plants of varying colors and leaves of different sizes to achieve whatever effect you want. One thing is certain, once your vertical wall garden has been built, it will surely become a topic of conversation whenever you entertain guests at home

Other Options For Vertical Gardens There are some other, easier alternatives for space-saving and realistic vertical gardens if all this seems like a lot of work to install and maintain.

Gutter garden

The gutter garden is super easy to install and can help you make the most of an unused or unattractive wall for growing flowers or even vegetables. Only add the old guttering rows to the ground, fill them with soil, and use them to plant whatever you choose. If you are going for this option, you must make sure you leave enough distance between each row, or your plants will not have enough space to grow. To keep soil from becoming waterlogged, you must also drill holes in the guttering for drainage. Also, keep in mind that you need to water them manually – so don't build them so high that you can touch them!

Pallet tray garden

Creating carton and pallet furniture is something of a fad right now, and you can just as easily turn pallets into vertical miniature gardens. Only turn the pallet on its side, and build plywood plant boxes. You may decide to mount it to a wall, so it doesn't tip over – and then just fill it with plants. It will be visually appealing and cost you next to nothing.

Clay pot vertical garden

Another appealing, simple to build vertical garden option can be produced by suspending clay pots on hanging steel rods. The highly original and perfect way to get your limited space to the fullest. It will take a bit of planning and preparation for implementation, but the outcome will be with the effort.

Wall trellis garden

Trellis has long been a traditional garden feature, but it can be given a modern touch with a bit of creativity and turned into a trendy vertical garden. One of these looks particularly striking when it is built indoors and is filled with plants. Even better, they are very easy to create, and very little financial outlay will be needed.

4.4 Best Plants to Grow in Vertical Gardening

Appropriate plant selection plays a significant role in the design and operation of vertical living wall gardens. In this post, we'll show you a collection of the best vertical gardening plants.

Ferns:

They are among the plants in the garden that are chosen for their adaptability and resistance to humidity. Ferns are easy to grow, and quickly cover the field. You can grow sword fern, blue star fern, bird's nest fern; it's easiest to grow. Generally speaking, ferns will grow downwards, and you will also need to grow other plants that cover with them.

Bromeliads:

Most bromeliads have shallow roots, and this makes them ideal plants for vertical gardens and needs little space to grow. A pleasant addition to your vertical garden may be their colorful leaves and long-lasting flowers.

Birds nest fern:

There is yet another fern that lives in containers effectively, needing the same conditions as the Boston fern. The yellow-green leaves in a wall composition provide a vivid contrast to darker greens.

Pothos, golden pothos, devil's ivy:

With good cause, it is a classic houseplant all over the world. It is one of the indoor plants that are more tolerant, tolerating low-light or bright sun, rain or dry spells, rich soils or poor soils, acidic, alkaline, or acid climate. It is difficult to get it wrong. It is a good choice for a vertical garden as a trailing vine because it can be encouraged to spread and fill any gaps.

Lipstick plant:

This is a lovely cascading plant that is often found in hanging baskets and is very suitable for vertical gardening. It will lead down and ascend. From its spreading stems and dark green leaves, vivid red flowers emerge, blooming sporadically year-round to a spectacular effect on a wall.

Succulents:

Sp. Sempervivum This is a hardy succulent, as its Latin name implies 'still alive.' Sempervivum is a succulent alpine, used to eke out a life between rocks and crevices.

Dracaena:

Dracaenas are common home plants, and most grow upright with strap-like leaves. Sometimes the leaves are filled with whites, creams, yellows, and reds.

Crotons:

Vivid shrub with leathery leaves, Croton is the most vivid in bright light. New leaves should be smaller in low-light conditions and pigmented less deeply. Grow Croton with high humidity at 60 to 85 degrees F.

Hosta:

Hosta requires shade, and cool weather. During summer, they have striking green, patterned varied leaves and white and lavender floral spray.

Baby's tears:

This is a lovely ground covered with small, bright green leaves that offer a walled garden with a delicate and soft appearance. It will grow in bright light as long as the sun is not too harsh, and the plant is watered. Baby's tears prefer well-drained soil in a temperate climate, with a variety of pH levels.

4.5 Maintenance of Vertical Garden

It's easier to maintain a vertical garden than a conventional garden plot, but that doesn't make it fully maintenance-free. Learn about caring for vertical gardens like watering, fertilizing, pruning, management of pests and weeds, and more. Plus, I'll give you loads of quick tips that make up a snap-in vertical garden maintenance. One of the key benefits of rising vertically is that it simplifies the maintenance. Growing up and off the ground plants make it easier to take care of them, and protect them from many rising problems. While it is simpler to look after the vertical gardens, they are not completely maintenance-free. But don't worry, there are plenty of things you can do to relieve the stress of maintaining the vertical garden. I'll tell you about vertical garden, so you don't have to waste the entire summer weeding, watering, to fertilize, and pest and disease prevention.

What form of upkeep is needed in vertical gardening?

We all know that one of the greatest advantages of vertical gardening is that it makes maintenance of a safe and efficient garden much easier. But what kind of vertical maintenance is actually needed in the garden?

Okay, when caring for vertical gardens, the activities you need to perform aren't so different. You're always going to have to wash, fertilize, prune and harvest – those don't go anywhere. But there are plenty of easy shortcuts you can take in a vertical garden that encourages such tasks. You would also need to be mindful of such issues as weeds, rodents, and diseases. But when plants are grown off the ground, many of these growing problems can be managed more efficiently, or eliminated together.

How to maintain a vertical garden?

Below are most popular vertical garden maintenance and care activities, and give you loads of tips to make them as painless as possible.

Watering

If your vertical garden grows in the grass, in the patio, or hangs on a wall, the same simple watering rules apply. Believe it or not, watering plants is the best way and a wrong way to go. Here are some tips on watering.

Water at the root

Watering plants at the root rather than above the top can help avoid the growth of fungus and mildew. It also helps to minimize weeds in a vertical garden plot, since only the plants are watered, and not all weeds.

Infrequent, deep watering

Watering your vertical garden less frequently is better, with deeper watering than watering it a little bit daily. Watering deeply helps the roots to grow deep, and the plants do not need to be watered as much as possible.

Provide sufficient drainage

Take particular note that all your vertical plantings and hanging pots have drainage holes to avoid overwatering. Without proper drainage, the soil soon becomes water-saturated, and the plants drown.

Mulch your garden

Mulching your garden has many benefits, and one of them is water retention. Mulch serves as an insulator, so the moisture in the hot sun does not evaporate. You may apply any form of vertical garden mulch over the soil, including those in containers.

Using irrigation systems

When it comes to watering, adding irrigation to your garden is a massive lifesaver. You might simply use your plot to weave soaker hoses or add drip irrigation. Plug your irrigation system into a wireless timer for the hose once mounted to make watering a snap.

Training & Tying

Several types of wine plants are outstanding climbers, and they can easily cover a trellis without your assistance. But, to grow vertically, some may need to be trained or attached to the support. There are some vertical garden maintenance tips for training and tying plants.

Training Vines

You can smoothly weave the vines into the trellis for plants that have tendrils or twining stems. Or you might tie them to the structure before

They catch them on alone. If the vines clasp the trellis, the links can be removed and, if necessary, reused on the support higher on new growth

Tying long branches

Plants with long, pliable branches would have to remain attached to the support to stay in place. Push the stem gently against the support to train these types of plants, and tie it to use twine, garden twist ties or flexible plant links. You do not want the links to strangle or cut through the vines as they grow thicker in either case, make sure to tie them very loosely. You may use plant clips, which clip on very loosely to prevent damage if you are concerned. Learn more here on how to train vines on a trellis.

Weed Control

When you garden in containerized vertical structures such as wall pockets, towers, or living art, the struggle to deal with weeds is practically inexistent. But looking after vertical gardens that rise in the earth, it's a whole different game of the ball. To control weed it is one of the biggest challenges faced by gardeners, and it can easily turn into an exhausting chore. But if you take a few quick measures to prevent weeds from growing in the first place, it doesn't have to be, so time-consuming.

Mulch

A dense layer of mulch is the perfect way to avoid weeds sprouting in your vertical garden. Growing winemaking crops on trellises and other structures for easy mulching around the plant's base. Place a 3"- to 4 "layer over the top of the soil for better

Performance.

Add a weed Barrie

Lay a thick layer of newspaper and a cardboard over the soil for an additional barrier, then spray it down before piling on the mulch. This smothered current weeds and gave you the upper hand. (I don't recommend using weed fabric though. Weeds will still grow on top of it, so it's really hard to deal with as your garden matures)

Don't water the weeds

Good irrigation would also help deter growing weeds. Watering the whole garden waters the weeds too with an overhead sprinkler. And then, specifically, concentrate the water at the base of each plant.

Fertilizing

Fertilization is an essential aspect of the care of the vertical garden. Many plant types can do their best in the growing season when they are fed regularly. It is very important when it comes to caring for vertical gardens growing in pots or hanging containers (such as living walls and baskets). They depend solely on us to provide the nutrients they need to survive. Here are some tips for fertilizing any kind of vertical garden.

Avoid chemical fertilizers

Avoid using synthetic pesticides while feeding any kind of vertical garden. Chemical fertilizers give us immediate gratification, but over time do serious harm to the soil's health. Burning the plant roots with chemical fertilizers, too, is much simpler.

Use organic fertilizer

By using organic fertilizers, you build up the soil to be a rich nutrient source for your plants. These days tons of natural fertilizers are available. This can either be added as a liquid (such as compost tea or natural kelp-based fertilizer) or as slow-release granules you apply to the soil.

Disease Control

Many disease problems begin at the soil level, so vertical gardens appear to have less soil-borne disease problems. You'll probably have to deal with some disease or fungus problems at some stage, though. So search for any signs of contamination, such as discoloration or spots on the leaves, as you go about your daily vertical garden maintenance activities. This way, before it spreads, you can take quick action to get the issue ahead.

Here are tips in your vertical garden for managing the fungus and disease issues.

Daily pruning

Pruning is the best way to avoid disease and fungus problems. Daily pruning allows sufficient flow of air and keeps the leaves off the ground.

Mulch

Mulching around the plant base in your vertical garden can prevent soil from splashing up on the leaves. This helps you prevent infecting your plants with soil-borne diseases.

Proper watering

Morning water plants so that the leaves will dry out throughout the day. This will help slow down or even stop disease transmission. Take care to keep water off the leaves if you decide to water your garden in the evening.

Proper disposal

Never put sick plants in your compost bin. Instead, dump the contaminated plant material into the garbage, or burn it to kill the pathogenic diseases

New potting soil

Never reuse potting soil or hanging gardens in your vertical containers. When planting vertical gardens in any kind of container or hanging planter, it is necessary to always use new, sterile soil.

Pest Control

The harmful bugs and fuzzy rodents are a part of nature for better or worse. But growing plants vertically makes it much easier to manage many of those pests. In the case of furry pests, vertical gardens that grow high above the ground are simply out of control. And climbing plants can easily be covered, growing in the garden. You can either encircle chicken wire or garden fencing at the base of fragile plants or create a fence around your entire plot.

Yet bugs can be a bit more difficult part of vertical garden treatment. Here are some tips for managing them.

Using physical methods

It is easy to manage most forms of insect pests using physical methods. Seek to pick them by hand, install row coverings or simply knock the pests off the plant with a powerful water spray from the garden hose.

Attract beneficial bugs

Many of the bugs in your garden are beneficial predators that prey on harmful insects, such as wasps and spiders. Plant extra flowers to deter such predatory insects, and aid with the upkeep of the vertical garden.

Remove toxic pesticides

If your prized plants are mumming with the bad bugs, it is tempting to search for the nearest pesticide to destroy them. Yet synthetic pesticides are not only detrimental to our well-being, but they are also highly harmful to the environment too. Plus, along with the poor bugs, they will destroy the good bugs right away.

Sparingly use organic pesticides

Organic insecticides (such as neem oil and diatomaceous earth) are a much safer choice. But even organic pesticides, when you find a pest insect invasion in your vertical garden, should not be the first thing that you search for. Just use them when it's absolutely necessary.

Target only the plague insect

It is important to use with extreme caution all forms of pesticides, including organic ones. Never apply any kind of insecticide big in your yard. Otherwise, you could end up destroying the good bugs when trying to rid your garden of the harmful plagues.

Pruning & Pinching

During the growing season, several forms of vertical garden plants can benefit from pruning or pinching. And getting into the pruning habit as part of your daily vertical garden maintenance routine is a smart idea. Daily pruning and pinching help keep the plants safe, monitor their size, and keeps them looking their best. Here are some short tips.

Deadhead flowers

Deadheading (pinching off the spent flowers) fosters new flowers on several plant forms.

Daily pruning

Daily pruning increases air circulation, avoiding problems with disease and the fungus. You can also trim the unruly vines to regulate their bulk.

Pinch to preserve the shape

To keep them small, Bushy plants can be pinched. Vining plants may be pinched or pruned to train them to grow over a large base, rather than growing longer and taller.

Remove dead or sick leaves

Periodically inspect your vertical garden and remove any dead or sick leaves. This will help prevent or even avoid the spread of mildew and disease, and it will also make your garden look the best.

Sterilize your pruners

Make sure your pruners are disinfected by cleaning them with soapy water, or by dipping them into rubbing alcohol after some diseased material has been sliced. In this way, you won't kill the other plants by mistake.

Winterizing

There's much of the time no extra vertical garden work you'll need to do in the fall. Most structures may just sit outside throughout the year.

However, if any of your vertical gardens grow in portable containers, then there are a few items to think about.

Save them inside

To prolong the life of your portable vertical structures such as picture frames, tower gardens, and hanging planters, place them in a shed or garage during the winter.

Overwinter containerized perennials

If, in any of your portable vertical gardens, you grow perennial plants, move the containers into an unheated garage or shed. It will provide them with extra security, and they have a better chance of surviving winter.

Dump the soil out

Clean up the container in the fall to cover your vertical planters. As it freezes, the soil left in the pot over the winter spreads, which can cause fragile containers to split or crack

Harvesting

When you cultivate vegetables upright, you'll get the added bonus of enjoying your harvest bounty. Here are some fast harvesting tips to add to your daily vertical garden maintenance schedule.

Harvesting often

Frequent harvesting gives the plant the energy to grow more food. So, test them daily for new vegetables ready to harvest once the plants are mature. The more you harvest, the greater the amount of food you receive.

Bigger isn't necessarily better

When left on the vine too long, some crops can get tough or seedy. So harvest your vegetables as soon as they mature to make sure you get the best flavor and the tenderest berries.

Remove damaged fruit

Damaged vegetables should be immediately harvested and either discarded (if sick or rotted) or consumed within one or two days.

It is much easier to look after vertical gardens than with a conventional garden plot but not completely hands-off. Follow these tips on vertical garden maintenance, and you won't have to waste your summer lugging watering pots, pulling nasty weeds, or battling pests and disease!

Conclusion

The term urban refers simply to the densely populated region or area which possesses the characteristics of the man-made environment. If you're a beginner gardener looking to jump-start your experience in the urban garden with far more success. You find a lot of detailed knowledge about growing a particular plant step by step in strong contrast to other gardening. Instead, It have taken the approach of explaining the why behind growing plants, giving you the basic knowledge to create your garden in an apartment in the area. It also tells you about container gardening and vertical gardening as well as the difference between urban and rural gardening. Phase by Phase Planting Details. This is perfect for the peoples who are living in an apartment and want to grow any garden in their apartment or want any fruits and vegetables for them; they can easily grow fruits and vegetables by reading this book and make a beautiful garden for themselves.

References

Urban Gardening - Definition From Ecolife.Com. [online] Available at: <http://204.236.232.61/define/urban-gardening.html>.

Organic Authority. 2020. What Is Urban Gardening? The Hot Trend That's Taking Over Cities. [online] Available at: <https://www.organicauthority.com/live-grow/what-is-urban-gardening-and-why-should-you-care>.

Admin, D., 2020. How To Plant An Urban Garden | DIY Home & Garden. [online] DIY Home & Garden. Available at: <https://diyhomegarden.blog/how-to-plant-an-urban-garden/>.

Now from Nationwide ®. 2020. How To Plant A Vegetable Garden In 10 Steps. [online] Available at: <https://blog.nationwide.com/tips-for-planting-garden/>.

EDGY_ Labs. 2020. Why Urban Farms Are The Future Of Food Production. [online] Available at: <https://edgy.app/urban-farms-are-the-future-of-food-production>.

Maximumyield.com. 2020. What Is Vertical Garden? - Definition From Maximumyield. [online] Available at: <https://www.maximumyield.com/definition/706/vertical-garden>.

Off The Grid News. 2020. 8 Advantages To Container Gardening - Off The Grid News. [online] Available at: <https://www.offthegridnews.com/food/8-advantages-to-container-gardening/>.

Blodgett, L., 2020. The Pros And Cons Of Container Planting. [online] Dailyimprovisations.com. Available at: <https://dailyimprovisations.com/the-pros-and-cons-of-container-planting/>.

Tirelli, G., 2020. Top 10 Benefits Of Living Green Walls - Ecobnb. [online] Ecobnb. Available at: <https://ecobnb.com/blog/2019/04/living-green-walls-benefits/>.

Ambius.com. 2020. What Are Vertical Gardens? | Ambius. [online] Available at: <https://www.ambius.com/green-walls/what-are-vertical-gardens/>.

The Family Handyman. 2020. 10 Best Plants For Container Gardening. [online] Available at: <https://www.familyhandyman.com/garden/best-plants-for-container-gardening/>.

BBC Gardeners' World Magazine. 2020. Top 10 Plants For Containers. [online] Available at:
<https://www.gardenersworld.com/plants/top-10-plants-for-containers/>.

www.ingramcontent.com/pod-product-compliance
Lightning Source LLC
Chambersburg PA
CBHW072012110526
44592CB00012B/1278